TREE FARM BUSINESS MANAGEMENT

TREE FARM
BUSINESS MANAGEMENT

JAMES M. VARDAMAN

James M. Vardaman & Co., Inc.
Forest Management Specialists

SECOND EDITION

A WILEY – INTERSCIENCE PUBLICATION

John Wiley & Sons · New York · Chichester · Brisbane · Toronto

Library of Congress Cataloging in Publication Data:

Vardaman, James M.
  Tree farm business management.

  "A Wiley-Interscience publication."
  Includes index.
    1. Tree farms—Management.   2.   Forest management.
I.   Title.
SD393.V25 1978   658'.93'49   78-1610
ISBN 0-471-07263-X

Printed in the United States of America

10 9 8 7 6 5 4 3 2

To Virginia
mother of my children, midwife of this book.

Preface

Tree farming is a great and rewarding enterprise because it deals with most of America's natural resources. Foremost are the trees themselves. As a natural resource, trees have a great advantage; they are renewable, not exhausted in use, and the cheapest known source of cellulose. Since the beginning of time, they have served mankind for fuel, shelter, and food, and even today many trees are used in their natural condition, merely cut or shaped into more convenient forms. This is changing as wood chemistry, reacting to the stimulus of an abundant and cheap natural resource, has developed paper, rayon, boards in which fibers are first separated and then reunited, charcoal, molasses, and other products in which the tree itself is unrecognizable.

Tree farms also contain minerals. Minerals have created fortunes in the most spectacular fashion, and the commercial possibilities of some minerals may be as yet unsuspected. Prior to World War II, how many of us knew the value of uranium? The basis for tree farming is the land itself. In addition to producing trees, however, land may produce crops of other kinds and provide the physical foundation for cities as yet unplanned. Tree farms also produce water, essential to human life and economic development and sometimes more valuable than any other crop. Mankind's knowledge is growing by leaps and bounds, so fast that scientists can hardly catalog it, and the impact of this knowledge upon natural resources has been and will be tremendous. Consider the technological achievements of the past twenty years. If this progress continues, what wonders the future must contain! Tree farms, composed as they are of many natural resources, must play a dominant part in this future, and tree farmers may ultimately harvest a crop that exceeds our greatest expectations.

But enjoyment of the things that constitute tree farming today depends on profits, and this book is an effort to help you make money from trees. It is written for the private tree farmer whose ownership is small and for the investor who hopes to become a tree farmer. In tree farming, ownerships of as much as 100,000 acres and representing assets worth millions are not uncommon, but our principle interest here is the farm of less than 5,000 acres, growing trees for sale on the open market.

There are 4,500,000 timberland owners, and they own nearly 500,000,000 acres. Obviously, it is impossible to consider many individual situations. All that can really be done here is to point out common problems, discuss their possible consequences, recommend the seeking of professional advice, and suggest how this may help you. The professionals whose services will be especially recommended are the consulting forester, the lawyer, and the accountant—all of whom can contribute directly to the health of your enterprise.

I hope the reader of this book will learn something of the many satisfactions and rewards that can be realized from tree farming. If he or she does, most of the credit should go to those who helped me prepare either the first edition or the second or both. I am deeply grateful to John D. Guy, Jr., Robert D. Hatcher, David R. Litterst, and other members of the management group at James M. Vardaman & Co., Inc.; the late J. E. McCaffrey, former vice-president of International Paper Co.; William M. Deavours, attorney; Woodrow M. Rogers, certified public accountant; Bruce M. Ferguson, consulting forester; Arthur W. Nelson, Jr., tree farmer; Professor Albert C. Worrell, Yale University; C. DeWitt Shy, consulting forester; Howard C. Ross, Jr., attorney; Professor John A. Zivnuska, University of California; Julius M. Ridgway, geologist; James E. Bryan, Jr., consulting forester; Donald E. Jameson, real estate broker; Dr. Alan C. Page, consulting forester; J. Mark Wilkinson, Prudential Insurance Co.; John W. Robinson, Jr., insurance broker; J. D. Strange, consulting forester; Everarde E. Jones, investment counselor; Robert B. Steele, certified public accountant; William K. Condrell, attorney; the late R. R. Covell, soil scientist; officers of Ridgway Management, Inc.; Paul B. King, banker; Billy Gaddis, state forester; Don Healy, Boise Cascade Corporation; Robert E. Wolf, Congressional Research Service; Jack H. Ewing, attorney; M. E. Benoit, Engelhard Minerals and Chemicals Corporation; Robert Edwards, Martin Marietta Aggregates; Charles F. Raper, Travelers Insurance Co.; Bill J. Morris, Federal Land Bank; and Albert L. Fairley, Jr., Hollinger Mines, Ltd., Toronto, Canada. I also thank Dr. James G. Yoho, John S. Tyler, Ben A. Davis, Jr., Stephen J. Doyle, Robert C. Hynson, Jason N. Kutack, V. B. MacNaughton, Jr., J. Walter Myers, James S. Roland, the late Admiral James K. Vardaman, Jr., Dr. Bruce Zobel, and Dr. Peter Koch.

JAMES M. VARDAMAN

Jackson, Mississippi
April 1978

Contents

1

What Is Tree Farming?

Tree farming is adventure. First, it is a new and pioneering business. Land and trees have been with us always, but harnessing them into production of periodic crops emerged as a business shortly after World War II. Although we have made tremendous strides, many mysteries of growing things are yet unsolved, and massive research facilities are exploring the unknowns of tree farming. Impressive scientific discoveries, however, may spring full-grown from the mind of one individual, perhaps your mind; by observation and study, for instance, you may discover that one of your trees is far superior in one characteristic to all others in the world. Second, it is adventure since none can predict the future, and your guess is apt to be as good as anybody's. For example, many believe that the future of the South lies in pine trees and that scrubby oak trees found in pine stands are weeds; you may disagree and have the last laugh if some scientist discovers that blackjack oaks produce rare chemicals that can be processed into the wonder drugs of tomorrow. Third, it is high adventure if you have retained some of the child's wonder at nature. All of us have a little of Daniel Boone in us; the wonderful world of nature fascinates us, and we soak up strength from direct contact with animals, birds, flowers, and trees.

Tree farming is individualism; indeed, I am convinced that the individual is by far the most efficient tree farmer. Although he fulfills his duties as a good citizen, he insists upon his right to act independently, and therein lie his power and his efficiency. Sitting on the throne of his tree farm, he regards the world with a critical eye and requires everyone to demonstrate continually the value of his services. He decides what use he will make of his own resources, what forest-management practices are adopted, who carries them out, when and how and to whom his timber is sold. He reserves the right to do nothing at all and even to dispose of his property if it fails to serve him properly. His personal ability is often the only limit to what he can do;

1

what his tree farm will produce is up to him and nature alone. He pays for his mistakes, but reaps the rewards.

Tree farming is fun. Can you explore the mysteries of life on a leisurely walk with your son through a factory? Can your daughter pick flowers in a bank? Do you choose an apartment house for a picnic? Can the bell of a cash register match the flute of a wood thrush? What other investment offers the appeal of a walk in the woods during a crisp autumn?

Tree farming is contentment. Hunger for land, even one small city lot, something on which to build the castle in which he would be king, is one of man's universal characteristics. A tree farmer standing in the middle of his 40 acres is lord of all he surveys. Of course, he cannot see very far, but he owns what he can see. Some of the poorest people own tree farms; their forest management is bad beyond belief, but they will almost do without food to preserve their ownership. Owning land satisfies a craving deep inside them, and who is to say that their land is not well used?

Tree farming is complexity. Biology and its forestry branch, law, geology, accounting, business management, salesmanship, economics, and politics are intertwined in the business of growing trees for profit, and the tree farmer ignores any of these fields at his own peril. Fortunately, however, he need not master all; specialists stand ready to serve him, and he must learn only to use them.

Tree farming is frugality and patience; the first allows time for the second. God makes trees slowly. Growing a tree from seed to minimum merchantable size often requires at least 15 years, and much longer periods are usually necessary for maximum financial returns; therefore, the tree farmer is necessarily a patient man. During the long wait, frugality, essential to maintain his ownership, is a day-to-day thing, for a wasted dollar now will be a large sum 20 years from now. Tree farmers are stable citizens, or their tree farms pass into other hands, and they must have the faith to plant crops that will be gathered by their children or grandchildren. What better base on which to build a nation?

Finally, tree farming is foresight. The long wait from seed to market keeps the tree farmer's eyes trained on the future. Fortunately, this forward look can be filled with anticipation, for right judgments can make him wealthy. Farmers raise a crop in 60 to 120 days; tree farmers raise a crop in 30 to 120 years. Who knows what kind, size, and shape of tree will be needed 20 years from now? The prescient tree farmer arrives at this market with exactly the right product and prospers by doing so. Uncertainty about the future, however, is no excuse for inaction; there are things he can do now to prepare for any eventuality, and wrong guesses may serve only to limit his profit to a modest one. Idle capital, or the failure to use productive resources, is the danger in this business; an abundant raw material always finds markets.

Defining tree farming as we have brings to mind two other questions—what is a tree farmer? and how does he go about his business? A tree farmer is essentially a capitalist, a manager of capital. Some capitalists measure their assets in number of shares or square feet of rental space or units of production capacity; he measures his assets in acres, trees, minerals, or inches of rainfall. Although his assets may differ from theirs in form, his function is that of all capitalists; he must manage what he has so that it produces a return commensurate with the amount invested and the risk he assumes. How does he do this? Primarily or exclusively by making decisions.

His decisions are many. In the beginning, he chooses his advisers—one in each of the fields of forestry, accounting, and law—and with their help he analyzes the business before him. First, he takes inventory in a complete fashion; he must know as much as possible about his assets and their capabilities. Next, he explores all possibilities open to him, giving due weight to each factor and deciding which assumptions and predictions are reasonable and which programs fit into his master plan. Then, convinced that tree farming has merit for him, he decides which programs to adopt and how to execute them. Finally, he brings the programs up for review at periodic intervals and decides whether to continue, reduce, or eliminate them, or to dispose of the assets entirely. In all these activities, he sits in the seat of power; he is the decision maker, and others carry out his orders. Few tree farmers are technically competent to carry out the fieldwork, and even fewer have the time or inclination to do so. Their fortes are management and sueprvision. Nevertheless, to be effective as managers, they must be familiar with the operations of the specialists at their command and the problems that face them, and, beginning with the next chapter, we shall examine these problems.

Although I have tried to make each chapter complete in itself, this is difficult when dealing with a complex subject, and you must consider the book as a whole and refer from one chapter to another when necessary. I thought about helping you with cross-references, but the footnotes became so numerous and irritating that I decided to leave you to your own devices. Throughout the book, you may find unfamiliar technical terms. I have not defined them as they occur, since this would break continuity and since both definition and discussion are sometimes necessary. You will find these terms in the Glossary, where I have also included terms that, although not mentioned here, are common in tree farming. I worried a lot about the order of chapters and finally decided that there is no logical order, since the situation of each reader differs slightly from all others. You will find that every important business aspect of tree farming is covered somewhere.

Now let us discuss this fascinating business.

2

Available Forestry Services

The woods are full of foresters who are available to assist you, and the great number of choices is somewhat confusing. These men vary widely in competence and experience, areas we cannot explore, but it is helpful to know what they do, why they do it, and who pays for them. Some are free and some cost money. First, let us discuss those who are apparently free.

FORESTERS EMPLOYED BY STATES

For many years, state governments have been concerned about the consequences of poor forest management within their boundaries and have adopted many programs to improve the situation. In many cases, programs are initiated by appropriations of the national legislature offering federal money to states, on a cost-sharing basis, for use in the cause of advancing forestry, and states are usually quick to take advantage of this opportunity. As the value of each program is demonstrated, states bear more and more of the cost, and the federal share decreases until it is often insignificant.

State forestry departments almost everywhere are responsible for fire control and use 80 to 90% of their budgets for this purpose. Tree farming without fire control is impossible, and you should support every move to increase the effectiveness of these programs. Money for fire control comes from many sources—federal and state appropriations, direct taxes on timberland, contributions by individuals or corporations. In some special situations, fire control is primarily handled by industrial associations. Taxes or contributions based on timberland ownership are fair, since beneficiaries pay in proportion to protection received, and needed increases in these assessments often meet with little resistance.

State fire-control equipment is idle much of the year and is usually available at modest rates per hour or day for plowing firebreaks on private land during the off-season. In an effort to make maximum use of personnel, some states have also allowed fire-control crews, during idle hours, to plant trees or perform timber stand improvement (TSI) work on a contract basis for private tree farmers. Such work during the fire season has not been entirely successful; it is almost impossible to coordinate firefighting with anything else, since fires do not start on schedule. Fire control is the major responsibility, and trying to combine it with other activities is like asking a city fire department to sweep the streets in its spare time.

Tree farming is best sold by demonstrations on the ground, and nearly every state provides some means for this. State forestry departments are usually organized with a forester for each county, parish, or district, and these men are allowed to spend a small part of their time in assisting private tree farmers with services and advice. They attempt to arouse interest in good forest management and turn the tree farmer over to private foresters once his interest is aroused and work of a substantial nature is required. The primary job of most state area foresters is fire control, and educational or promotional activities must not interfere with the main effort. As forestry grew in importance, functions of state forestry departments expanded. State forests, school lands, state parks, and other state lands came under their management, and foresters concerned with actual management of tree farms, either public or private, appeared at state and district headquarters. These now perform many services offered by private foresters; charges, if any, are set by state headquarters.

When these programs began many years ago, tree farmers, knowing little about forestry, appeared unwilling to pay for advice and services. Therefore, most services were free but were limited in some way, either in number of days worked or in number of acres marked. Soon there was dissatisfaction with the relative inefficacy of free work, and now the situation varies from state to state. Most offer advice free; some offer nothing but free services, while some charge for all services, and some charge for certain services and offer others free.

States are usually the main source of tree seedlings, since development of idle land is in the public interest and large-scale operations are necessary to reduce cost. To provide an incentive to tree farmers, some states offer free seedlings in limited quantities and sell additional quantities at cost. The federal government encouraged seedling production in the beginning by agreeing to bear half of any operating losses, but federal contributions are now very small in most cases.

Recently, many states have become interested in markets for tree farm products, especially where there are surpluses, and have added foresters specializing in wood utilization to the state staff. These men promote the establishment of additional manufacturing plants, assist existing plants with technical manufacturing problems, and promote the use of wood by working with architects and builders. Such programs have just begun, and their effectiveness is yet to be determined. They benefit you indirectly by increasing markets for your products.

The state extension forester and his staff usually operate from the state agricultural university, and he does not perform on-the-ground services. His function is primarily education, and his concern is the wide dissemination of the latest forestry information. He can help you only in a general way. He is paid by the state, but part of his salary comes from federal subsidies.

Your state forestry department will be happy to describe its program to you if you will call state headquarters or the area forester near you.

FORESTERS EMPLOYED BY THE FEDERAL GOVERNMENT

You will have little direct contact with a federal forester who can help you in actual management of your tree farm. As you have just seen, the federal government exerts its powerful influence on private forest management by indirect activities in fire control, nursery operation, and extension and management work. Federal funds are also available for cooperative work with states and individuals in detection and control of forest insects and diseases. Foresters employed by the federal government supervise these programs closely to make sure that money is used in accordance with the desires of Congress.

You might have direct contact with the federal government through subsidies for forest-management practices disbursed through the Agricultural Stabilization and Conservation Service (ASCS). This money is allocated by an ASCS committee in each county, and each tree farmer must apply for funds to the county committee. Amount of money available for various practices varies widely by county from year to year; you can find out how much you can get by calling your local office.

Although this money appears to be free and can be a great help, you should weigh certain factors before asking for it. First, the amount of money you can get may be so limited and the procedures for getting it so burdensome that the request is not worthwhile. Second, you cannot get any ASCS money unless the work done on the ground meets ASCS specifications, and you might wind up spending more out-of-pocket money in trying to qualify for "free" money. For example, the ASCS-approved plan for establishing trees on a given area might call for hand planting of pine seedlings at $40 per acre; since the ASCS would pay 75% of the cost, your out-of-pocket expense would be $10 per acre. If the area could be regenerated by broadcasting treated pine seed at a cost of $5 per acre, you would save money by avoiding the ASCS program. (See page 96 for a discussion of problems that might arise from the tax treatment of government subsidies.)

Please remember that this is just an example; planting rates and costs and the ASCS policy may not apply to your area. All foresters connected with the ASCS program realize the great difficulty in tailoring a nationwide program to fit perfectly each individual situation, and they constantly strive to eliminate waste and poor forest-management practices. Nevertheless, until they solve the problem completely, you should examine the local ASCS program carefully before entering it.

For many years, the federal government has been the leading, or only, agency in forestry research, and tree farming today owes much of its success to this research. It is carried on in offices and field laboratories at U.S. Forest Experiment Stations in every timbered section and covers every aspect of tree farming. Although the emphasis is on biology, increasing attention is directed toward financial aspects. Published research reports are available at little or no cost and can be valuable to you. They may be obtained from U.S. Forest Experiment Stations, and any forester can give you the proper addresses. Under certain conditions, the station near you will put you on the mailing list for every publication issued, and you might find this helpful if you have the time and ability to absorb such volumes of information.

An exception to the rule of indirect contact is a case similar to the Yazoo–Little Tallahatchie Flood Prevention Project in northern Mississippi administered by the U.S. Soil Conservation Service. Hundreds of millions of pine seedlings have been planted under direct supervision of foresters of the U.S. Forest Service, who work closely and directly with tree farmers of the area. So far, effort has been concentrated on tree planting and TSI work for the primary purpose of controlling erosion, but, as the trees reach merchantable size, increasing attention is being directed to management. The forestry program of the Tennessee Valley Authority began in a similar fashion but has evolved mainly into forestry research and promotion. The use you might make of such a program depends entirely on the local situation.

Most U.S. foresters you will see are engaged in management of national forests. They are busy with major responsibilities in managing these large areas and are not permitted to assist private landowners. They can still be helpful, since they have many of the same problems you have. They sell timber on the open market and pay to have various forest-management activities carried out. They usually have recent stereoscopic aerial photographs of the national forests, and these may cover your land if it is nearby. They are familiar with everything that goes on inside the forest boundary and will be glad to keep you informed of developments that might affect your land. They travel roads through the forest constantly and may pass through your land once a week. You should go out of your way to meet the district ranger stationed near you.

FORESTERS EMPLOYED BY INDUSTRY

Forest-products manufacturers have always helped tree farmers because they depend on them for most of their raw material. In the early 1940s, many companies hired "conservation foresters" to work with tree farmers on a free basis to improve woods practices and promote forestry. These men often marked timber, advised on forest management, distributed free seedlings, planted trees, and performed TSI work at little or no cost.

In the mid-1970s these efforts developed into what are generally called landowner-assistance programs, although each company selects a unique name for its program. Typical programs start with a written agreement that the company will carry out most forest management activities either free or at cost and that the landowner will pay the taxes, maintain the boundary lines, and give the company "first refusal of all forest products sold from the land." Agreements can usually be canceled on short notice. The apparently innocent words enclosed in quotation marks above effectively eliminate competitive bidding on timber sold from the tract. You will discover what this does to the landowner by reading Chapter 4.

Industrial foresters are active in the work of forestry associations, and this is a highly beneficial trend for all concerned, especially you. Although association work is time-consuming and often thankless, the accomplishments are most valuable. Individual tree farmers have a great stake in this work, but often cannot spend the necessary time on it. State forestry associations fight for appropriations for state forestry departments, oppose legislation harmful to tree farmers, offer rewards for conviction of woods burners, promote the use of all wood products, study ad valorem taxes on tree farms and other common problems, and, in general, serve as organized voices of forestry. Industry associations have similar programs but cover a wider area and concentrate on problems of the particular industry. For example, lumber industry associations advertise frequently in national magazines to promote the use of lumber; this expands markets of lumber companies and, thereby, markets for products you grow. All associations are active in the fight to retain capital gains tax treatment of tree farm income. Industrial foresters participate in these activities because their employers benefit directly and substantially. The fortunate thing about it is that you have many of the same goals, which you probably could not reach without this great help.

MISCELLANEOUS FREE SERVICES

Much of America is organized into soil conservation districts, legal subdivisions that receive technical assistance from the U.S. Soil Conservation

US Dept of Agriculture
US Soil Conservation Service
form - management plan

QUANTITY	AUTHOR AND SHORT TITLE			PRICE
1	VARDAMAN TREE FARM 2 ED	2640	2701B	17.50
	SBN* 0471072635X			

PACKING
LIST

NET WEIGHT   1LBS 1OZS

NET AMOUNT
SALES TAX
SHIPPING AND HANDLING CHARG
CASH RECVD WITH ORDER AT E

TOTAL
QUANTITY
SHIPPED

1   K 72847

PLEASE ADVISE WITHIN 30 DAYS
OF ANY DISCREPANCY
JOHN WILEY & SONS, INC.
ONE WILEY DRIVE, SOMERSET, N. J. 08873

Service and other agencies; these districts are concerned with wise land use, which sometimes includes forestry. For no cost, local employees will prepare a farm-management plan that includes a soil map of your tree farm. You can determine what services are offered in your area by calling the local office; it is listed in the telephone book under United States Government, Department of Agriculture.

All land-grant colleges conduct valuable research through agricultural experiment stations, and some of this applies to forestry. Your state may have a forestry school where even more information is available. Schools conduct forestry research concentrated on state problems, and many professors do part-time consulting work on a fee basis. Research results are usually published, and many projects can be inspected under certain conditions. Any forester in your area can help you reach the proper persons.

FREE SERVICES VERSUS FEE SERVICES

A familiar axiom states, "Free advice is worth what you pay for it." To a certain extent, the saying applies to forestry. The big trouble is that the men described above work directly for somebody else and indirectly for you. This leads to several complications.

First, their employers find it difficult to judge the quality of their fieldwork. These services are usually offered in addition to the main effort of the company or agency, and attention must be concentrated elsewhere. The tree farmer served is seldom qualified to judge how good or poor the work is and rarely has any direct contact with the employer. Furthermore, these foresters often have many other responsibilities and can hardly be expected to be experts in everything. For example, a state forester may spend 80% of his time on fire control, 10% on administration, and 10% on management services to tree farmers. He may be mediocre in management but so outstanding in fire control that he is properly classified as a valuable employee. His ability as a forest manager is not judged in daily competition. Although none of these employers will knowingly permit poor performance, each finds it hard to specialize in everything.

Second, their experience may not cover many areas of vital importance to you. With the possible exception of industrial foresters, they are primarily experts in biology and know little about the business aspects of tree farming. For example, few of them are familiar with the details of ad valorem taxes, one of the largest management expenses. Biology produces trees, but business management produces profits.

Third, their loyalty must be divided since they are paid by somebody else. This division sometimes puts a conscientious man in an uncomfortable position. Consider the case of a paper-company forester marking a stand for a pulpwood thinning. He must please both the tree farm owner and the pulpwood buyer, who is a dealer for his employer. Trees of many sizes can be manufactured into pulpwood, and some larger trees, which are more profitable for pulpwood producers, may also be suitable for poles and saw-timber. How many of these should he mark? Conflicting interests of the two principals make any decision subject to criticism. Especially during the first selective cut, many large, open-grown trees, primarily suited for pulpwood, should be cut ruthlessly; by doing so, however, he may be unjustly charged with favoring the buyer when his treatment is exactly what the seller needs. Many men working in this field are competent and experienced, and all of them are conscientiously trying to do a big and much needed job. The difficulty lies in the nature of their employment; they must be on both sides of the trade.

Men who provide forestry advice and services for fees do not have this trouble. They are responsible to the man who pays them, and the relationship is clear cut and more comfortable for all concerned. Although being in private practice does not guarantee greater competence or better ethics, it does subject them to the test of competition every day. If their services are not valuable, they will soon be bankrupt, and this is as it should be. Since there are no arbitrary limits on services that can be performed, their experience tends to be wider and more closely related to problems of the tree farmer. There is no division of loyalty, and often their work is checked by both the tree farmer and some of the free foresters described above. The most important private practitioner is the consulting forester, but there are others who might serve you well.

NON-PROFESSIONAL FEE SERVICES

In general, nonprofessionals have little formal education or training in forestry; they have learned by doing. Nevertheless, they may be very capable in their specialties and can be valuable to you.

A special class of contractors has grown up around the ASCS program of federal subsidies, and they are usually known by the ASCS term *vendor*. They plant trees and perform TSI work on a contract basis and are familiar with the details of doing this under the ASCS program. They are often farmers who keep busy at this during winter months. They use farm labor of the area and can handle jobs of considerable size. They quickly absorb lessons learned from foresters who inspect their work, and they take great pride in good performance. They live in the area and benefit from observation of the work they have done in prior years. You can locate these men through the local office of the ASCS and investigate their individual qualifications by talking to foresters who have used them. Large industrial tree farmers and the U.S. Forest Service use contractors who perform the same services. These contractors tend to be bigger than the ones described above and may be better qualified and organized. You can contact them through the men who use them. There are some important legal aspects to your relationship with contractors, and you must investigate this.

Before the rise of consulting foresters, many tree farms were managed by what we shall call "timbermen," for want of a better term, and some of them still practice, although their number is dwindling. Many have spent their entire lives in the woods—scaling logs, estimating and buying timber, logging, surveying lines, and performing all other tasks connected with tree farm operation. They are often competent, reliable, honest men who are eminently practical. They are probably not familiar with all business aspects of tree farming, but they are unbeatable in the woods. These men are uncommon and hard to find, and I can suggest only that you inquire for them among foresters in your area.

HOW CAN YOU USE THIS TALENT?

Such an array of services is apt to be bewildering, and understandably so, and I am not able to tell each of you individually how to use them. Nevertheless, certain guidelines are clear from what I have said before, and certain other points become plain from long experience in the field.

First, the most important single thing you can get from any forester or agency is good advice, and I emphasize "good." Advice comes in a flood from all of the people described above, and all of it is offered in a sincere effort to help you. On the other hand, it should be clear that quality varies widely. I suggest that you choose a man of wide and long experience and unquestioned reputation among professionals in the field and that you stick with him even when he tells you things that are unpleasant. He must be able, as near as possible, to see things from your exact standpoint. This would seem to exclude everyone except consulting foresters, but this is not invariably so. Some of the finest advisers are not in consulting work at all, or practice on a nonprofessional basis, or may be entirely outside the field of forestry. I am convinced that a good adviser is a hundred times more important than all other elements put together. This is true whether you own 40 or 400,000 acres.

Second, and secondary in importance, you must pick capable people to execute programs set up by the adviser. Both functions may be combined in the same man and often are, but you might be able to economize by using several people. There are several possible solutions to your problem, and they depend mainly on size of ownership.

Suppose you own 160 acres, by most standards a small tree farm. Nevertheless, it represents a value of about $16,000 or more, and $16,000 is much more than pocket change. Such an investment requires good management. For advice you might turn to a consulting forester near the tree farm. During a year, you will probably use no more than one whole day of his time. Several people may have a part in actual operations. Your tree farm inventory might be taken by a consulting forester or local timberman. Data on soils might be obtained from the Soil Conservation Service or a soil scientist in private practice. Boundary lines might be established by a timberman or county surveyor. Tree planting and TSI work might be done by a contractor with or without ASCS subsidy. Inspections to prevent trespass and to watch for important changes in the timber stand might be conducted by you personally, one of your neighbors, a consulting forester, or a timberman. Timber sales should be handled by a qualified authority. This program leaves out most free services, because few of them can help you in the actual work. Research reports issued by various agencies described above will advance your general knowledge of forestry. You should compare notes with any foresters you contact, since this is a form of business intelligence work; you might pick up some information useful to you or your adviser.

More foresters and politicians worry about and study you than all other tree farmers combined, and they have tried many schemes to improve the quality of your forestry practices. Often, they have decided that tree farming cannot be profitable for you, because your land area is too small to afford good management, and they have used this as a justification for government subsidies and programs with the subtle controls that may accompany them. I believe this approach is wrong; I believe that tree farming can be very profitable for you and, indeed, that you may be more efficient than any other class of tree farmers. All you have to do is face economic facts with the expert help available.

If you own 1600 acres, your problem is similar, except that you probably cut timber every year instead of every five or ten years. I suggest the same scheme of management. The quality of your adviser and the amount of time you spend on the property become more important because you have a sizable asset.

If you own 16,000 to 160,000 acres, you have a substantial business that involves operations on a daily basis. Tree farms of this size are large investments and have special management problems not easily solved even today. You need advice of the highest caliber and services in considerable amounts. Many such tree farmers have groped for a solution to the management problem by hiring a forester shortly after his graduation, with no qualifications other than a good education and some practical work during his college summers. If this is better than no solution, it is a poor one. The young and inexperienced forester is overcome by the complexity of his task. His head is crammed with information, but he knows little about how to apply what he has learned in school and learns every day on the job. His education was short on business, law, and accounting, and the forest-management principles, which seemed so simple and logical at school, are suddenly difficult to apply. At the same time, he costs money. His salary, transportation, equipment, and related costs amount to $16,000 yearly at a minimum, and this is 40¢ per acre on a tract as large as 40,000 acres. A more subtle problem arises because you hired him. His mind belongs to you, and you tend to do what he recommends, since you are paying him. Instead of guiding, however, he needs to be guided; he may be headed for a brilliant future, but he needs help from an experienced hand along the way. After several years experience with an arrangement of this kind, the tree farmer realizes that he has made a bad mistake that cannot be corrected for years.

Necessary services such as timber marking, planting, TSI work, and boundary-line maintenance constitute a major part of the problem. Although the minimum amount of work required is large, it varies from year to year and with the season of the year. Full-time employees may be fully utilized at some times and almost idle at others, and there is always the temptation to use them, merely because they are on the payroll, for projects that are not essential. In addition, their travel time to work may be excessive if the land is scattered, and there are few chances at such economies as might come about when adjoining landowners maintain the same boundary line. These housekeeping chores represent important management costs and can be most cheaply performed when ownership is consolidated or when independent contractors perform them for many landowners in the area. Such contractors achieve the benefits of consolidation without the necessity for land transfers, and they might be the proper solution to this problem.

As always, the most important part of the problem is the need for advice, and the correct solution depends mainly on total land area. In the lower part of the size range, outside advice is almost essential. The caliber of advice needed makes it prohibitively expensive to hire an adviser on a full-time basis. You probably do not need advice more than two or three times each quarter, but it must be excellent when you get it. As total land area increases, it is more and more likely that a full-time adviser is needed. Such an adviser may be highly profitable when combined with the use of contractors for housekeeping chores. He can supervise the contractors from time to time to ensure that his recommendations are carried out and spend the majority of his time considering the problems and possibilities of the enterprise.

Some tree farm cooperatives exist, and more are suggested, but you can see from discussion here and elsewhere in this book that the idea has no advantages for you unless ownerships are contiguous or unless the total area managed by the cooperative is large—large enough to hire the best brains. Even under these conditions, cooperatives reduce your freedom of action (and complete freedom of action is the great advantage of individual ownership) and may result in more expensive operation. You should not become part of one without serious thought.

3

The Consulting Forester

The most important professional man to you is the consulting forester. His special talents, knowledge, and experience eventually bring him into contact with every tree farmer when decisions of consequence are involved. Because of his great importance, you should use particular care in selecting him.

SELECTION OF THE CONSULTING FORESTER

The words "consulting forester" are found in any dictionary, and they are presently used of men whose talents, experience, and character vary widely. A list of those practicing in each state can be secured from the state forestry department. Many are members of the Society of American Foresters; many are licensed or registered by several states; some belong to the Association of Consulting Foresters. Although these agencies strive for higher standards, none have effective police powers. None of these memberships guarantee the same level of competence you might expect from someone with the title of doctor, lawyer, or certified public accountant. Most consulting foresters are graduates of a college-level forestry school, but some excellent ones are not. Many practice in the highest professional tradition and can make tree farming profitable for you. What qualifications must a good one have, and how can you investigate him?

First, a consulting forester must be competent. You can determine his competence by asking the people who deal with him in his daily business. Talk to his clients and to several people who know him from long business experience with him. His banker and his wife will usually speak kindly of him, and so will members of his church and civic club, but their opinions attest more to his popularity or other personal attributes than to his competence in forestry. Talk to several pulpwood buyers, paper companies, and sawmill operators in his area; they know how good or bad he is.

Second, he must be above reproach in his business dealings. His methods of handling the timber and money of his clients must be clear, careful, and straightforward and must compare favorably even with operations of the trust department of your bank. In particular, payment for timber should always be made directly from buyer to seller and should pass through accounts of the consulting forester only in those rare cases when he is acting in some capacity under a court decree or other legal arrangement. His clients of several years' standing can tell you about this.

Third, he must be independent. Under no circumstances should he be engaged in buying any forest products whatsoever. Some men using the title of consulting forester are primarily pulpwood dealers; this means that they tell you what your timber is worth and then buy it from you. In my opinion, a lawyer cannot be on both sides of a lawsuit, and a consulting forester cannot be on both sides of a timber trade. The same advice applies to those who seek to manage your tree farm merely for the privilege of buying timer that comes from it. This is a poor solution to the problem of forest management and should be discarded as soon as possible. It is hard to believe that you can get something for nothing, and good business requires a man to count his own money.

Fourth, he must be engaged in his business on a full-time basis. The kind of service and advice you need can come only from the man who gains his experience and knowledge by constant and diligent application. You need a man who understands your problems thoroughly and who is solving similar ones every day of his business life. They are too important to be solved on a part-time basis.

Fifth, he must be thoroughly familiar with the important aspects of your local situation. He must also be aware of all the complexities of this business and know when to recommend other professional advisers. Using information presented elsewhere in this book, question him anout several problems peculiar to your land. You will be able to tell shortly how well he knows his business.

Sixth, he must be articulate. He deals with terms, measures, and tools that are highly technical, and the details of his operations may be unintelligible to many tree farmers. These technicalities are not beyond the realm of human understanding, however, and a competent consulting forester can explain them to you regardless of how little you know about the business of tree farming. If he cannot explain them to you, he probably does not understand them himself.

SERVICES OF A CONSULTING FORESTER

Consulting foresters perform every service needed by the tree farmer and can be used profitably on even the smallest area. Most are what you might call "general practitioners," but highly trained specialists are just as common in this professional field as in any other. You should realize this, for most tree farmers need a general practitioner, not a specialist.

The general practitioner is usually extremely competent at estimating timber volumes. He takes inventories for the tree farmer, calculates growth, and sets up a forest-management plan exactly tailored to the needs of his client. He marks timber for cutting, estimates its volume, sells it, and supervises its harvest. He may also take timber inventories for manufacturing concerns that desire to purchase timber from land not under his management. He is well versed in the personality, character, and qualifications of everyone in the timber business near him. To a certain extent, he specializes in these generalities.

He also plants trees, performs TSI work, and establishes boundary lines (although he is probably not a surveyor). He often appraises damage done to timberland by pipelines, power lines, highways, fire, trespass, mineral operations, and so forth, and he usually makes an excellent witness in a lawsuit where technical forestry information has an important bearing on the case. He appraises estates for inheritance-tax purposes, especially where values are likely to come under searching examination by the Internal Revenue Service. He is especially good at resource surveys for forest products industries seeking new locations. Existing timber volumes can be obtained from published reports, but the important questions are who owns the timber? can it be bought? and what are the prices and conditions? These data are revealed by his daily business. His independent appraisals are almost essential for tree farmers seeking loans. Only rarely does he have anything to do with fire suppression. Nevertheless, he is experienced in fire control (he must have such experience to practice forestry) and often prepares fire-control plans for all tree farms under his management. He may have heavy equipment to plow fire lines and perform controlled burning, but this is unusual.

In certain cases (usually involving properties in excess of 10,000 acres), he is responsible for the entire operation of a tree farm. He assesses the land, pays taxes, collects rentals on leases of cultivable land, supervises recreational use, and handles all other administrative details, in addition to the main forestry operations. He supervises the property to prevent trespass and continually advises the owner about forest-management practices that should be undertaken.

Occasionally he sells land for his clients, but he usually does this only when his technical competence is necessary to present the property adequately. Most consulting foresters do not do well when the sale of a property depends heavily on romance. He sometimes buys land for his clients but rarely for himself; purchases for his own account may put him in competition with his clients.

He has many clients, sometimes hundreds, both large and small, and no client pays him more than a small fraction of his annual income. This is a healthy situation for all concerned. It contributes greatly to his independence and enables him to give truly unbiased advice. He values each client but is able to be uncompromising when matters of principle or technical competence are involved.

Specialists are what you might call "foresters' foresters." They may be experts in aerial-photograph interpretation, forest soils, tree diseases, genetics, continuous forest inventory using the most advanced computers, statistics, game management, or any of many other fields. The general practitioner may need specialists from time to time, and he will inform you of this necessity if it arises. You need a general practitioner, not a specialist. You must realize that the term *forester* can mean many things, and you should investigate thoroughly before employing anyone.

CUSTOMARY FEE ARRANGEMENTS

The fee basis among consulting foresters is as varied as in any other professional field, and I can give you only a general idea of what to expect. He will welcome your inquiry on this subject, since it is wise to get this matter settled early.

He makes inventories, growth studies, and forest-management plans at a quoted price per acre. He handles all details of a timber sale on a commission basis, although the agreement with him will probably provide some compensation for his cost if the timber is not sold. His experience with timber sales enables him to predict almost exactly what the sale will produce, so this provision seldom applies. In some cases, either he or his client prefers that he work on timber sales on a per diem basis; this is a matter of choice. He plants trees and performs TSI work on some sort of contractual basis and maintains boundary lines at a fixed charge per mile. He does almost all appraisal work, and testifies in court when necessary, at a stated fee per day and expenses. He sells land on the same basis as he sells timber, but it is almost impossible to sell land without a recent and reliable tree farm inventory, and he may require one. For office conferences and advice, his charges are based on the amount of time used.

When he is responsible for the entire management of a tree farm, each case is a special situation. Management agreements usually provide for a small monthly or quarterly retainer with additional fees based on amount of timber produced or other work done. Although there is usually a written memorandum or letter covering these situations, most consulting foresters have found that a long-term forest-management contract is a waste of time. These contracts must contain adequate escape clauses for both parties, and neither forester nor client wants to continue an unsatisfactory relationship. Every consulting forester knows that the most binding agreement is one that has grown by custom through years of mutually happy relations. Such agreements have no fixed life. They may be terminated by either party on little notice, but they usually endure for many years.

His fees are generally deductible expenses for you, but some must be capitalized, and some reduce the size of capital gains. For instance, fees for supervising planting operations must be capitalized, and commissions on timber sales reduce the size of the sale. Your tax adviser can settle doubts about tax treatment of his fees.

Consulting foresters are relatively new in this country, and they are enthusiastic about their profession and love to talk about it to people with a genuine interest. Even before you hire one, you can discuss your problems with him without fear of trespassing on his time. He will let you know gently long before you reach this point, and you will find that he is anxious to help you.

ADVANTAGES AND DISADVANTAGES

The consulting forester has one disadvantage; he costs money. It appears, however, that he is worth it. Although competition soon eliminates nonessentials, and every part of the timber business is competitive, the number of consulting foresters continues to increase.

He has one big advantage. He is on your side. Growing trees for profit is a highly technical and long-range business, and you need the best possible assistance in order to succeed. If you make a mistake in accounting, you can find and correct it in a matter of hours. If you make a mistake in tree farming, it may take you several years to realize it and 20 years to correct it. A consulting forester who meets the qualifications mentioned above offers you a valuable opportunity. Any forest management costs money. The best is reasonable, and the cheapest often ruinous.

4
Sale of Timber

No step in tree farming is more important than the sale of timber. Sucess of the whole operation depends on income from sales, and yet, for most tree farmers, sales come only every 10 or 15 years and, for some, sales come only once in a lifetime. You can spend 50 years growing a crop of timber and then lose much of it in 50 minutes by mistakes in selling. Selling is the key operation; you cannot spend too much time studying how to do it. You should start by considering whether the product you have for sale is scarce or plentiful.

THE MYTH OF TIMBER SCARCITY

Timber is *not* scarce. No matter what you have heard, we are not running out of timber. But acting as if we were can reduce your profit almost to nothing. The things you would do if timber were scarce are ridiculous if it is not.

How can we tell that timber is not scarce? My firm gets this message loud and clear from the timber market, for we sell timber for landowners every day.

Government reports tell the same story. U.S. Forest Service inventories of timber by states have been showing for years a surplus of growth over removals in nearly every eastern state. Although some types of trees are less plentiful in certain areas and although our forests are clearly not producing all they could even with our present limited knowledge, the picture is definitely one of increasing abundance.

A 1975 study sheds additional light. In "Is Timber Scarce? The Economics of a Renewable Resource," Dr. Lloyd C. Irland of the Yale University School of Forestry states, "No *general* timber scarcity has existed in this country since 1950. . . . . Recent decades have brought persistent declines in real prices of major forest products. Our balance of foreign trade in timber products has steadily improved since 1950. And despite rising consumption of industrial wood, the quantity, quality, and accessibility of our forest resources have been steadily improving for the last 30 years. . . . It appears that the United States enjoys a timber-producing capacity equal to the needs of a population double its present size."

Therefore, you are not dealing with desperate buyers who will fight to get what you have. You are competing with a horde of sellers who want the same dollars you are after. When any product is plentiful, it must be marketed with great skill if you want a good price for it. The first step in your marketing is to learn the mechanics of the timber market.

HOW THE TIMBER MARKET WORKS

No market I know of is like the timber market. A phone call can get you a firm price on many common items: stocks, bonds, groceries, clothing, commodities, and so on. But a phone call to 20 timber buyers will likely get you 20 different estimates, and each will want to see your timber before making a firm price.

Many forces cause the timber market to be so entirely different. Some are obvious to all; other powerful ones are known only to a few. You must, first, understand how complex the market is and, second, learn how to cope with this complexity.

The first step is to discover who the buyers are and what they want. There are many. One list for Mississippi alone contains 462 names. Each uses trees of some sort, and all buy on the open market.

Each, in turn, manufactures the trees he buys into products for his market, and there are many markets. To illustrate the situation, here are the 16 buyers in one county and what they want:

Buyer	Wants Trees That Can Be Made Into
A	small-size pine lumber
B	pulpwood and logs of all species
C	pulpwood
D	pine lumber
E	pulpwood
F	pine and hardwood logs
G	all forest products
H	small-size pine lumber
I	all forest products except poles
J	tree-length pulpwood
K	pine plywood

L   pulpwood pine and hardwood lumber

M   pine poles and piling

N   hardwood veneer

O   pulpwood and poles and piling

P   pine and hardwood lumber

Some of these buy from and sell to each other at times, and each also has a hookup with buyers of forest products that he does not want.

This is the situation today, but it changes constantly. Some buyers will disappear and be replaced by others. Others will change their requirements; paper companies, for example, formerly bought only pulpwood, but now want everything.

Our next step is to determine where these buyers are. Timber can be hauled economically about 75 miles from woods to mill. Therefore, the buyers listed want timber in several other counties, and buyers at many other places want timber in this county.

Each buyer has limits on the kind of tree he can use. For example, a pulpwood buyer can use anything that can be reduced to chips; a plywood buyer can use only fair-quality trees large enough to be peeled on a lathe; a pole buyer can use only trees of a certain size, taper, species, and straightness. But, when you examine the whole spectrum of desires, you can see that almost all can use a reasonably straight 12-in. pine tree.

These 12-in. trees, acceptable to many, are worth quite different amounts to each, however, and the difference depends on what the buyer plans to make from the tree. Such a tree on today's market brings about $1.90 for pulpwood, $4.50 for saw-timber, $5.70 for plywood, and $7.30 for poles.

I say "today's market" because prices are never stable, and markets for manufactured trees are independent of each other. Pulpwood demand fluctuates with demand for paper, which fluctuates with general business conditions; sawtimber demand fluctuates with demand for housing; plywood demand fluctuates with demand for many products; pole demand fluctuates with level of utility construction.

Another factor, unrelated to those just discussed, also causes wide differences in prices. So far as we know, no buyer owns enough land to grow all the timber he needs. Therefore, all must buy much of their raw material on the open market. Although there are many sellers and although the supply is greater than the demand, an open-market buyer can never know where all his timber will come from or what each tract will cost. He knows only that his average manufacturing costs are so much and his average selling prices are so much, so his average timber cost must be less than the difference between them.

Therefore, he starts each year trying to buy timber as cheaply as possible.

First he looks for distress sales (rights of way that must be cleared quickly, damage from windstorms or insects) where timber is cheapest. Next he negotiates directly with timber owners, many of whom know little about the timber market. Next he posts notice that he will buy logs of certain specifications delivered to his mill and pay so much per MBF. Next he tries to negotiate long-term cutting contracts in which the prices he must pay are set according to a published index of prices, preferably of the product he manufactures.

Each tactic produces some timber. Over the course of a year, results might be as follows:

% of Requirements	Method of Buying	Price MBF	Total Price
10	Distress sales	$50	$ 500
30	Negotiations	60	1800
20	Delivered logs	75	1500
20	Cutting contracts	80	1600
80			5400

His *average* cost for 80% of his needs is thus $5400 ÷ 80 = $67.50/MBF. But his mill will not be profitable if he runs only 80% of the time. His fixed costs must be spread over a larger production, and his workers cannot earn a decent living by working only 80% of the time. Therefore, he cannot stop buying here. But, since these tactics have succeeded so well, he has a lot of room for maneuver. If he can pay an average of $100/MBF and still make a profit, arithmetic shows that he can pay as much as $230/MBF for the last 20%.

This is a hypothetical case. I put these figures in merely to illustrate my point. But something like this happens to every timber buyer every year. All buy timber by every means they can think of and pay a different price every time.

To get the remaining 20%, our buyer must now turn to the professional market. Here competition is much stiffer. Not only are sellers much better at their business, but also he encounters other buyers whose situations are like his. He cannot tell in advance what they will do, and he may not get a second chance to buy what he needs. Therefore, his only alternative is to make his best offer the first time.

Perhaps you can now see that the timber market is huge and extremely varied. There is no "market price" in the sense that there is a market price for General Motors stock. On most trading days, the high for General Motors is not 5% above the low, whereas the high bid for a tract of timber often exceeds the low by 100%. Furthermore, the market is changing all the time, and this means that timing of sales is important.

TIMING

Compared to other kinds of farming, tree farming has a unique advantage. If you do not like the market this year, you can postpone sales, and conditions may improve. Even if there is no improvement, growth during the waiting period will increase the size of the harvest. Therefore, you have some leeway about when to sell, and good timing can pay handsome dividends. One aspect of timing is the attempt to sell at market peaks. No one knows whether present market prices will go up or down or stay the same. If your sales are infrequent, however, you must make some decision, and there are several people who can help you.

First, there are timber buyers. They are in the market every day and can give you good information on prices. They can give you more than opinions; they can offer you money you can spend. As your timber becomes more desirable, you will see more and more of them as they call upon you in efforts to buy it. When the market is good and timber is hard to buy, these visits become more frequent, perhaps bothersome, but this is a sign of excellent markets. The visits go the other way when the market is poor.

Second, there are other private tree farmers, some of whom operate large areas on a full-time basis. They sell timber of all kinds almost every day and have done so for years. They are constantly comparing notes with other tree farmers and will welcome a chance to talk shop with you. You should look for someone with about 5000 to 10,000 acres who is not engaged in forest-products manufacturing. You should also realize that he is an expert only in his particular situation, and his may not be the same as yours.

Third, there is the consulting forester. He makes most of his living from timber sales, and probably no one else knows as much about the price of as many different forest products.

Fourth, there are those foresters with the U.S. Forest Service and state forestry departments who work directly in management of federal and state forests; the district ranger on a U.S. national forest near you is a good example. They often sell timber and have personal knowledge of market conditions. Personal knowledge is far better than information passed around in the organization.

Fifth, there is your state forestry association. All persons and companies interested in forestry belong to it, and it is an excellent source of information about present and future markets. Many of the private conversations at meetings are on markets. Memberships are available at moderate cost and provide data on all forestry subjects. Additional information is available from the Forest Farmers Association and national and local associations of manufacturers.

If your timber sales occur at frequent intervals, this aspect of timing is not so vital, since the law of averages works for you. Another aspect of timing now becomes more important, namely, that of selling each tree when it is no longer a profitable investment.

Value of an individual tree as an investment changes constantly throughout its life. In general, its annual percentage increase in value is extremely high just after it reaches merchantable size, and it declines gradually as the tree grows larger, until the rate eventually dips below that commonly paid on savings accounts. Therefore, within certain limits, a young tree is a highly profitable investment, and an old one should be liquidated. For instance, a 6-in. pulpwood tree, by growing 4 in. in diameter, often increases in value 433%, and this growth may take place in as few as 12 years. On the other hand, a 16-in. sawtimber tree, by growing 4 in. in diameter, often increases in value only 71%, and this growth can take place in 12 years only under good forest management.

The first cut from your tree farm will probably be a selective one, and it goes without saying that you will remove diseased, crooked, forked, defective, and suppressed trees. Biology will select these for you. After all of these are cut, you may notice that the remaining trees are still too crowded. Which of these should be cut?

You can answer this question for each tree by a slight modification of the procedure described under "Growth Study" in Chapter 8. Measure the tree's diameter now, and calculate its present value. Then calculate its value when it becomes 2 in. larger. Divide the increase in value by the present value to get the percentage increase the tree will make by this growth. Now, take an increment core, and count the number of rings in the outside inch. This shows how many years it took the tree to put on the last 2 in. in diameter, a growth rate it will probably maintain in putting on the next 2 in. Using a compound-interest table, determine the annual rate that corresponds with the percentage increase and the number of years just obtained. Let us suppose that the tree has a value today of $2.00 and will be worth $3.00 when it is 2 in. larger and that the increment core shows eight rings in the outside inch. Therefore, the tree will increase in value 50% and will probably do so in eight years; the compound-interest table shows that its annual growth rate is 5.2%. If the interest rate you selected in your financial forecast is more than this, the tree should be cut.

Performing each step outlined above for each tree is an impossible task; fortunately, it is not necessary. If the timber marker performs this calculation a few times every morning and repeats it occasionally each day, he will find that the general appearance of the tree and its relation to its neighbors are reliable indicators of its growth rate and that he can predict almost exactly what an increment core will show.

You must use judgment in applying this procedure when you come to clumps of good trees so closely spaced that none are growing at a satisfactory rate. It is reasonable to assume that removing two trees from a crowded clump of six will speed up the growth of the remaining four to an adequate rate. Training and experience in biology are necessary to make sound judgments of this type; this explains why foresters spend such a long time in school and later study. You must also be sure that the trees left after cutting are sufficient to make adequate use of the soil's productive capacity. If they are not, it is usually time to liquidate everything and start again. Removal of less than 45% of the trees generally means that the remaining trees will expand to cover the ground, and removal of more than 75% of the trees probably means that the stand should be liquidated. The area between these guideposts requires expert advice. A partial cut is not always good; it can be bad business and biological practice. Let us move on to a discussion of how to go about selling timber once you have decided that the timing is right.

SEALED BIDS VERSUS NEGOTIATION

Selling timber by sealed bids is the superior method by far. It is mandatory in some cases and highly desirable in most others and has advantages for both buyer and seller.

The most obvious advantage to the seller is the higher price that usually results. The high bid is often appreciably above the second highest, and this is not possible at an auction. A well-prepared and properly advertised sale of this type definitely establishes the market price even if there is only one bid. The method also treats all potential purchasers fairly and impartially and, when properly conducted, virtually eliminates the suspicion that sometimes arises from private negotiations. It is most satisfactory for estates with scattered heirs of all ages. It eliminates negotiation on price, and, since the bids are written and sometimes accompanied by good-faith deposits, the buyer's commitment is firm.

Surprisingly, the system has some advantages for the buyer. Where it is well established, he is somewhat like a housewife at a supermarket. He receives a steady stream of sale announcements and can pick tracts he wants from a wide assortment. He saves money on timber inventories, since preparation of the sale is properly the seller's responsibility and includes an independent inventory. A prudent buyer can spot-check this without the expense of a complete inventory. He benefits from fair and impartial treatment of all prospective purchasers. He meets his competitors in this marketplace and checks the market prices of timber in his vicinity.

One disadvantage of the sealed-bid method is that it will not always work. This is especially true when large volumes are involved. The market is active where individual sale volume varies from $10,000 to $100,000, but the number of bidders decreases rapidly as amounts grow larger. Almost all sales involving hundreds of thousands of dollars must be negotiated. Negotiation is also the only way to sell poor timber, since demand for it is always weak. It is often the only solution when specifications for the forest product are varied and complicated, a situation quite common in selling poles and piling, oak-stave logs, walnut, dogwood, and so forth.

Popularity of the sealed-bid sale has increased rapidly with the growth of tree farming, but some owners are excellent traders and prefer negotiation. If you want to try this, remember that a timber buyer is a skilled professional who is rarely outtraded by an amateur.

LUMP-SUM SALES VERSUS PAY-AS-YOU-GO SALES

Sale for a lump sum is usually superior. It is clear, definite, and fast, permitting flexibility for tax purposes in most cases and eliminating much of the negotiation and argument that all too often accompany pay-as-you-go sales. Title passes immediately to the buyer, who then has a certain time to cut and remove the timber; this transfers to him the risk of loss by fire, insects, weather, theft, and so on. When combined with sealed bids, lump-sum sales render timber almost as merchantable as listed stocks. One frequent dispute in the sale of timber arises over how it will be measured, and this is often combined with fear that the timber will not be paid for. Lump-sum sales eliminate these problems.

You will be able to sell most of your timber for lump sums, but in several cases you will have to use the pay-as-you-go method. You will certainly want to do so in selling large, overmature timber where values per unit are high and where the presence of many defects makes it impossible to determine the volume of standing trees. You should do so when selling timber so inaccessible that no buyer will risk paying a lump sum for it except at bargain prices. You may have to do so to obtain capital gains treatment of the profit on the sale; your tax adviser can tell you. You will also have to use this method when the market is made up primarily of small operators, which is often the case in the sale of pulpwood. Many pulpwood buyers are capable men, but their available funds are just sufficient for trucks and other production equipment, and they have nothing left over with which to buy timber. Insistence upon lump-sum sales in this kind of market means that you must take a licking on price.

If you are forced to make pay-as-you-go sales, you should require the buyer to put up a deposit large enough to ensure that he will fulfill the contract. Twenty-five percent of the anticipated sale volume is common. The sale agreement should specify the timber to be cut and the time allowed for cutting in an unmistakable manner, and the buyer should forfeit his deposit if he fails to meet these conditions. The air is always full of sweet reasonableness and good intentions during negotiation of a contract, but they vanish when the hard work of logging is compounded by bad weather. Once the buyer starts logging, he should be compelled to finish, since you probably will not be able to get another to clean up the mess.

SEALED-BID *AND* LUMP-SUM SALES: A COMBINATION TO COPE WITH COMPLEXITY

As I pointed out earlier, the timber market is huge, extremely varied, and changing all the time. Nobody can ever know enough facts to figure out what it will do. Location of a buyer is easy to determine, but his hopes, fears, and present financial status can only be guessed at. Anyone trying to predict which buyer will be highest bidder on a given tract is sure to be surprised. Two examples will give you an idea of how analysis can fail.

Two recent sales were equal in area and pine volume. They were held nine days apart. Both contained trees suitable for a wide variety of products and were located in the same market area. We prepared both in the way described in the next section and sent invitations to bid to the same 125 buyers. We received these bids:

	Sale 1		Sale 2	
Bidder	Distance from Tract	Bid	Distance from Tract	Bid
Plywood Plant A	49 miles	$80,240	65 miles	$ —
Plywood Plant B	8 miles	77,124	35 miles	79,353
Plywood Plant C	73 miles	—	59 miles	84,910
Lumber Company A	50 miles	74,440	15 miles	80,151
Lumber Company B	63 miles	—	80 miles	86,656
Paper Company A	25 miles	69,308	30 miles	60,475
Pulpwood Dealer A	30 miles	66,334	60 miles	63,641
Pole Company A	45 miles	—	10 miles	92,812
Pole Company B	45 miles	—	10 miles	90,286

We were surprised at several results. Why was Plywood Plant A the high bidder on Sale 1 when it was 41 miles farther from the tract than Plant B? Why did Plywood Plant C fail to bid on Sale 1 and yet become a strong contender on Sale 2? How could Lumber Company B almost win Sale 2 and yet be so far away? How could Pulpwood Dealer A bid more on Sale 2 than Paper Company A when he was farther away and normally would sell the timber to Paper Company A?

The changing-all-the-time characteristic affects different buyers in such different ways that no single buyer can and will pay top dollar all the time. Each January for five straight years we bid off one block of timber from the same tract. Listed below are all bids for each year. As you can see, only two buyers bid every year, and although their five-year totals were higher than others, neither succeeded in buying any of the timber. No single buyer bid a total that was more than 85.7% of the highest total. Here are the bids:

Buyer	1976	1975	1974	1973	1972	Total
Highest	$89,331	$97,587 [a]	$228,432	$92,812	$101,705	$609,867
Buyer A	88,350	92,705	195,965	79,353	66,406	522,779
B	71,672	92,613	179,082	60,475	66,613	470,455
C	89,331	48,492	195,424		74,745	407,992
D		77,100	166,772	84,910		328,782
E			228,432	90,286		318,718
F	62,040		142,689	80,151		284,880
G		87,000	179,200			266,200
H			166,739		71,477	238,216
I			194,560			194,560
J		52,459			101,705	154,164
K		33,000		92,812		125,812
L				86,656		86,656
M	83,333					83,333
N		67,753				67,753
O				63,641		63,641
P					44,000	44,000
Q		10,034				10,034

[a] Sum of high bids on three separate parcels.

When results are so unpredictable, when you can never know enough to figure out what will happen, what can you do? There is only one solution to the problem: Make your sales both sealed-bid *and* lump-sum. The sealed-bid feature makes all buyers compete against each other. The lump-sum feature puts all of them on an equal footing and eliminates the need for understanding technicalities.

PREPARATION OF SALE

The most important thing in preparing a timber sale is to determine *exactly* what you are selling. I emphasize *exactly* because this is where many sellers fail. They estimate timber volume by measuring samples of the tract and then multiplying them by a factor to get a total. And, since woods work is hard and may be expensive, they spend a lot of time figuring out how to reduce the size of sample theoretically needed.

This fancy figuring is perfect in theory, but wrong in practice. Theorists forget that ordinary people measure the sample, that ordinary people make mistakes, and that, in practice, most mistakes cause an underestimate of the actual volume.

One common mistake in practice is to select a sample that is not representative of the whole. Try estimating how many people are in your house by counting those in one room, the kitchen, and then multiplying by the number of rooms. If everyone has gone to bed, you will estimate that the house is empty.

Another common mistake is to measure the sample incorrectly. I have worked in the woods for 30 years and know how easy it is to make mistakes when the weather is terrible or you are very tired or the tract is a huge briar patch or you step on a snake or get eaten up by mosquitoes.

Finally, for reasons we've never understood, mistakes in measuring a sample usually cause an underestimate. Perhaps timber cruisers think that the worst mistake is to overestimate volume and therefore choose a conservative course when they have a choice. Perhaps it is easier to miss a tree on the sample than to include one *outside* it. At any rate, we know from long experience that by far the largest number of mistakes cause an underestimate and that this error can easily be 25%.

The way to avoid these mistakes is to measure 100% of the trees (not a sample) and measure them with a steel tape (not a Biltmore stick). This requires a lot more work. But you can pay for it by finding 5% more timber, and you are likely to find 25% more.

The measurement should be made by a person of unquestioned competence, integrity, and independence. (For obvious reasons, most timber buyers will not bid on measurements made by a competitor, especially when the competitor might bid on the same sale or have refusal of the timber.) His report must show the number of trees by species, DBH classes, and lengths, and also his estimate of the volumes they contain. Even if buyers do not agree with his estimates, they can calculate their own from the tally of tree numbers.

The second step in preparing a sale is to make the terms as attractive as possible to buyers. You should certainly furnish them a copy of the report of the man who measured the trees. This makes it easy for them to decide whether they are interested and to check the figures, something they must do. They also know how accurate a 100% measurement is, and the reputation and ability of the measurer will go a long way toward convincing them that the volume is actually there.

You should also try to meet other terms buyers want. For example, although 12 months' time is usually long enough to allow for cutting many tracts, an unsettled economy or difficult logging conditions may cause buyers to lose interest unless they have 24 months. Severe restrictions on logging can also reduce value of a tract to zero, so you must strike a happy medium here. Access to the tract is essential; if you do not provide it, you force each buyer to work out his own and thereby reduce his interest by raising his cost of bidding. Finally, you must allow enough time for buyers to appraise the tract; three weeks is usually sufficient.

The third step is to advertise the sale thoroughly. You should send an invitation to bid to every buyer within 75 miles. In much of the eastern United States, a circle of this size will contain 125 buyers. If you do not have the names and addresses of this many, you should advertise the sale in classified sections of all newspapers in the area.

The invitation to bid should contain maps of the property that can be followed by timber buyers of reasonable competence. One with the scale common to most state road maps will help buyers locate the tract and the meeting place for the show-me trip. A more detailed, larger-scale map of the tract itself will help them greatly in their appraisal work.

The invitation should also contain such details as time allowed for cutting timber, penalties for cutting trees not included in the sale, instructions as to how to make and submit a bid (including date, place, and hour), a copy of the conveyance to be used (if buyers are not familiar with your sales), provision to allow seller to reject bids, and any special conditions of the sale that might be important to buyers. Some sellers require each bidder to attach to his bid a good-faith deposit in cash or equivalent, but this increases the cost of bidding and should be avoided if possible. Your forestry adviser can tell you if it is needed. You must provide a qualified man to show the timber to buyers and answer questions about it, and the invitation should contain the time and meeting place for such a show-me trip.

OPENING BIDS

The best strategy on bid-opening day is: "Operate in a gold-fish bowl." In the past some sellers, who hate to be cheated themselves, have figured that any trick played on a timber buyer was fair, so timber buyers have good reason to be suspicious of secrecy. The more open your conduct is, the more convinced they will be that they have received fair treatment, and the more anxious they will be to deal with you again.

You must adhere strictly to the time limit set forth in the invitation. No bids should be considered except those that arrive on time. Taking bids over the phone is acceptable in these days of poor mail service, but be sure they arrive before the deadline.

You should open all bids publicly and announce them. If no bidder is present or if some of the sellers are absent, you would be wise to have the bids opened by a third party, perhaps an officer of a nearby bank. Soon after the opening, you should tell all bidders about results of the sale.

Although the bid opening should be public, deliberations about the bids are quite often private, and everyone understands why. For example, the seller might want to consult his forestry or tax adviser about certain aspects of the sale, or approval from all members of a joint ownership may be required. Nevertheless, bids are like marriage proposals; they usually expire if not accepted quickly. If there must be a delay in acting on the bids, you should tell the successful bidder what your problem is and ask for a reasonable time to solve it.

Some sellers attempt to use a sealed-bid sale merely to establish a trading floor from which to start an auction after bids are opened. This practice infuriates timber buyers, and I do not blame them. It destroys the ability to make future sealed-bid sales and may destroy the sale at which it is attempted. If you do not like the bids, reject them all, and then do what you want. Everyone's customers deserve fair treatment.

CLOSING THE SALE

After the best bid has been accepted, you should prepare the conveyance and submit it to the buyer for his approval. By taking the initiative here, you can usually assure that the language in the conveyance is what you want, and this is often much quicker and easier than trying to rework a fill-in-the-blanks form used by many buyers. You cannot close the sale until the buyer approves the conveyance, but most buyers will go along with any reasonable form.

While this is going on, the buyer will ask his lawyer to make an independent check of your title. Although you do not intend to convey something you do not own, you may not know about some flaws in your title. Such an outside check, which is free, can help you find and cure them. This check usually takes three or four weeks.

You cannot get your money until the sale is closed, so there is not a moment to be wasted here. Many things can happen to upset the sale: a tornado may rip up the timber; the buyer may die or go bankrupt; one member of a joint ownership may die. You can prevent closing delays on your side by working with your lawyer before you get ready to sell. He can help you remove many obstacles.

INSPECTION OF THE LOGGING OPERATION

It is essential to supervise the sale until the last tree has been moved from the woods. All sale agreements should provide penalties for cutting trees not included in the sale, and these must be strictly enforced. The reason is obvious in lump-sum cash sales of marked timber. You should also watch logging operations in pay-as-you-go sales where some of the timber to be cut is relatively inaccessible. Many loggers want to cut what is easy to get and leave the rest. This is equally important in complete liquidations. It is expensive to harvest trees passed over in the main logging operation, and these scattered trees will impede development of the next crop. Larger buyers are responsible and ethical businessmen who will not countenance any violation of a sale agreement, but they cannot supervise every movement of every woods employee or contractor. Loggers and their employees are often short of money, and even today some timber strays into the hands of unauthorized persons. Proper supervision of logging reduces these problems to a minimum.

MISCELLANEOUS CONSIDERATIONS

Although it may be desirable from a forest-management standpoint, it is seldom possible to sell two different products such as sawtimber and pulpwood in the same sale. Two logging crews under different management on the same area at the same time make it almost impossible to establish responsibility for violations of the sale agreement, and create constant arguments about who has priority in the use of roads. These considerations make it poor business practice for you to reserve tops in a sawtimber sale, hoping to salvage some pulpwood from them. It is far better to put the responsibility for woods operations squarely on one buyer.

Much timber sold today is marked for selective cutting, and you can get a good idea of the ability of the timber marker by looking at the actual paint marks on trees to be cut. There are always at least two marks, one at eye level and one below stump height. An experienced marker puts all paint marks facing the same direction, uses plenty of paint in a bold and unmistakable manner, and is always careful to apply the bottom mark so that some paint sprays on litter at the stump. If your timber marker uses little dribbles of paint on first one side and then another, beware; you are probably in the hands of an amateur. An experienced timber marker often puts on so much paint that a casual observer thinks he has cut the stand too heavily. Do not be alarmed if this happens to you; it is a common optical illusion. The tendency is often to mark too few trees, and you have no worries if you are careful in selecting the timber marker.

A desirable sale involves trees with a large average volume, and you should keep this in mind when selecting areas to be cut, method of cutting, and future objectives in your forest-management plan. It is quite common for 10-in. pulpwood trees to sell for 20 to 40% more per cord than 6-in. trees. Difference in the size of the average stick causes a sizable decrease in logging cost with an increase in diameter, and every pulpwood producer and logging contractor is well acquainted with this fact. Foresters are anxious to thin a stand as soon as possible to speed up or maintain the percentage growth rate, and inexperience leads some to attempt thinning too early. It is sometimes better to suffer some crowding temporarily in order to get an increased price per unit of volume. Timing of thinnings is a complex matter and requires an experienced adviser.

You may hear some talk about managing stands on a pulpwood rotation or a sawlog rotation and wonder what all this means to you. The answer is practically nothing. The decision about this is of primary importance only to the tree farmer growing raw material for a manufacturing plant. Your main objective is money, and you have to maintain flexibility to reach it. Insofar as possible, you must be able to change your forest-management plan quickly and often, and it is not necessary, and may be a big mistake, to make fixed plans for 20 or more years into the future. Your task is to keep your productive acres busy growing something—reproduction, pulpwood, or sawtimber—and the only thing you need to fear is idle land that can be used.

It is difficult to sell a tract of timber by contracting with a logger to deliver logs to various markets. Determination of proper logging cost is a complex problem even for professionals. Logging depends greatly upon weather, and what starts out to be a job requiring one month sometimes requires one year. Log-storage capacity at each mill is limited and varies with the season of the year; therefore, log production must be coordinated with mill production. Log specifications vary from mill to mill, and it is often hard to see the subtle but significant differences among them. These considerations alone make a sale of this kind a management job of major proportions. In addition, the legal relationship between you and the logging contractor may not be clear, and this is dangerous in the event of accidents involving people or property. You should not attempt such sales without expert advice.

TIMBER-DAMAGE APPRAISAL

Related in many ways to the sale of timber is the destruction or damage of it. The party causing damage expects, and may be required, to pay for it, and determination of the proper dollar value is hard. Damaged timber is not often sold and may include such unsalable items as reproduction. If it is sold, the amount received is more likely to be its salvage value than its value under normal circumstances. In some cases, destruction is total, but it is often only partial. Forest fires kill some trees and reduce growth and resistance to decay of many others. Timber damage may be only part of the total damage to a tree farm; for instance, construction of a limited-access road will destroy timber on the right of way itself and may reduce the value of the remaining timber and land. Destructive agents such as salt water from oil wells may not only kill existing timber but also reduce or eliminate the productive capacity of the affected area for a variable period of time. Timber may be killed at a time when its owner would not consider selling, such as during temporary market dips or while trees are small and growing rapidly into the products of high value.

In nearly all cases, these factors make it necessary to substitute an expert appraiser for the market itself. His job requires thorough knowledge of both forestry and finance, and he must be able to explain his procedure in convincing terms, since he will often testify before a jury whose members are not tree farmers. Damage claims in thousands of dollars are common, and they may run into millions in large forest fires. If you are on either side of a damage claim, you should see your lawyer and, with his assistance, choose someone to appraise the damage. Although many feel competent to do this, most are not qualified by either training or experience, and few make good witnesses in a courtroom. Your lawyer probably has some experience with appraisers in this field and knows which men can be qualified as expert witnesses. In many cases such as those involving rights of way of all kinds, you will be able to determine much of the damage before it occurs, and most appraisers prefer to inspect the trees standing on the area to be cleared rather than to estimate what has already been removed by sampling an adjacent area.

Money received for timber damage gets much the same treatment as income from timber sales, but you should consult your tax adviser about this.

5
Miscellaneous Income

In a sense, the term *tree farm* is a misnomer. Trees are the primary product of this factory, and most income comes from their sale. Tree farming, however, is really brains and money applied to natural resources, and its major attraction is that it offers the investor many ways to make a profit. The average tree farm is a grab bag, and tree farmers were experts in multiple use before foresters discovered the term. Income from sources other than trees is always important and sometimes may provide spectacular profits. Let us consider some possibilities and the problems they present.

RIGHTS OF WAY

Rights of way (ROW's) now occupy millions of acres of our land. A casual glance at an aerial photograph of any of many areas reveals a striking pattern of strips occupied by roads, pipelines, power lines, highways, and railroads. Nearly every tree farmer has had some contact with ROW's or soon will have. The expanding population requires goods and services in enormous amounts, and many of these move to market over ROW's of one kind or another.

Sale of ROW's is one of the plus factors (see Chapter 9) that cannot be measured, and you need the profit offered by sales of this kind. The price per acre is often three or four times the market price of the land through which the ROW passes. This makes the sale attractive, but there are good reasons for this premium price. Although each ROW agreement must be investigated thoroughly with your lawyer, there are several considerations that apply to all.

First, ROW's often have a damaging effect on parts of the property outside the ROW itself. For example, they may obstruct free travel from one side to another. Second, they may increase administrative costs. It takes time to study and negotiate proposed ROW sales, and you are often left with an annual ad valorem tax burden on land you cannot use. Third, they may limit future chances for profit. A large pipeline ROW through your tree farm may greatly restrict its development for residential or industrial subdivisions. Fourth, certain phrases in the usual agreement may cloud the title to other parts of the property. Fifth, the usual agreement often contains conveyances of certain rights you may want to keep for future trading. Although some ROW's may increase the value of your tree farm, any ROW may have far-reaching effects, and proposed sales should be carefully studied. Let us look at certain factors that vary with the purpose for which an ROW is used.

Pipelines. Construction of pipelines usually involves burying the pipe and, consequently, disturbing the soil and its cover. In hilly country, subsequent erosion is possible. Gullies are obvious dangers, and you may also have trees adjacent to the ROW killed by heavy deposits of silt. The pipeline company has an interest in preventing erosion, since it may expose the pipe. The company's self-interest may protect you from this hazard, but it is far better to provide for erosion control in the ROW agreement.

Usual agreements prepared by companies grant them rights to construct additional pipelines on the property either on the present ROW, approximately parallel to it, or in unspecified places. Tree farmers commonly assume that additional lines will be laid on the existing ROW and as close to the original line as possible. Under certain agreements, however, the company may have the right to construct a second line at right angles to the first. The agreement may also grant the company rights of access across all land described therein. Such agreements may limit future development of the entire tract, and you should avoid them. These rights cannot be acquired by eminent-domain suits, and you should save them for future trading.

Buyers of ROW's usually offer a certain amount for the ROW plus an additional amount for damages done to the surface, such as destruction of timber. Total consideration is the important thing. You can usually settle damages either in advance or after construction is completed; the better alternative depends on the the situation, but you must realize two things: First, settlement of damages after construction delays payment, opens the door for interminable haggling, and may make a lawsuit necessary to determine the amount. Second, settlement of damages in advance requires that you sign a release. Be sure that the release covers only damage on the ROW itself. Additional damage may be done to adjoining portions of the property.

After much trading, you may find that the company will settle for a permanent ROW of narrow width for the pipeline itself plus use of an adjacent strip for the period of construction, which additional strip will revert to you after construction. You can often make this trade without reducing the amount of money you receive. This depends mainly on your trading ability;

the company can probably acquire a permanent easement over the entire width through the power of eminent domain.

Power lines.   The most important characteristic of power-line ROW's is the provision allowing the company to cut danger trees. Ordinarily, a danger tree is a tree standing outside the ROW that would hit, or come within 5 ft of hitting, a structure or conductor if it fell directly toward it. Such a provision is essential for the company; your only concern is to see that the agreement provides for payment for these trees. They will be cut as long as the power line is used, and it will be almost impossible to salvage them. Some old agreements may not contain this provision.

Roads and Highways.   Generally, construction of a new road through your tree farm adds value to it by improving access and increasing development possibilities. An exception to this rule is the limited-access road. Such roads cannot be crossed except at designated points, and their construction may eliminate access to certain portions of your tree farm. In addition to studying the agreement, you should make a thorough inspection of the proposed ROW on the ground. Minor logging barriers such as streams may become major ones after the road is constructed.

Highway departments and private companies have one thing in common: both try to acquire by agreement rights that they cannot acquire by eminent-domain suits. Highway departments are particularly anxious to obtain access over other parts of the tract in order to reduce construction costs. It is a good idea to treat everyone equally and never convey anything that cannot be taken from you unless the grantee is prepared to pay for it.

Many existing highway ROW's contain substantial amounts of merchantable timber, and ownership of this is often of interest to you when you harvest timber. Your lawyer can determine this. When the timber is on the ROW, the highway department probably controls its disposition, but it may have to distribute the proceeds of its sale to you.

Logging Roads.   Requests for logging ROW's are common and, in most cases, are granted without charge. Many tree farms do not touch a road used by the public, and the giving of logging ROW's to your landlocked neighbors may pay handsome dividends in goodwill. Someday you may want a logging ROW yourself. Such ROW's should be written and have a definite time limit. If properly prepared, they can be recorded and thereby eliminate easements by adverse possession. Details of this subject are too technical for this book; you should discuss them with your lawyer.

Trading Pointers.   The time to settle the difficulties involved in ROW's is before you sign the agreement, not after. Pressure is on the buyer, and you should keep it there until all details are settled in a satisfactory manner.

In particular, you should avoid provisions that permit future operations on your tree farm with only damages to be negotiated. You should insist that anyone desiring to enter your tree farm negotiate with you each time such entry is desired.

Occasionally, ROW's are used and later abandoned; in rare instances, ROW's are obtained and never used. Agreements should provide for reversion to you after use is discontinued or unless construction or use begins by a specified date.

In many forms, location of the ROW is described only in general terms, and indefinite language often leads to trouble by casting a cloud on the title of all land described in the agreement. The ROW should be surveyed and marked with stakes, and the agreement should contain an exact legal description of its location.

In many cases, an ROW to be cleared contains merchantable timber, and what to do with it is a subject of negotiation. As a general rule, it is wise to sell it to the ROW buyer and let him worry about it. Difficulties of salvaging it are many. The time allowed for salvage is limited; the total volume is relatively small; and the area on which it is located is a long narrow strip to which all salvage operations must be confined. These factors reduce prices substantially.

Some ROW's are easements only, and some are conveyances of land. If the ROW is an actual transfer of title, it can probably be removed from your ad valorem tax assessment. If it is an easement, you should leave it on your assessment, since you still own the land and the ROW may revert to you in later years. You have some grounds for a reduction in assessment in such cases, but tax assessors are hard to convince. The best solution is to see that the consideration for the ROW involves an amount on which interest will be sufficient to pay ad valorem taxes forever. Only your laywer is qualified to explain differences in easements and transfers of title.

Eminent-domain suits take time and cost money for all parties, and most agencies prefer to avoid them if possible. This fact may enhance your trading position. Because ROW purchasers are professional traders, it would be most unusual if their initial offer were their best. Inasmuch as the buyer knows that the law will always require him to pay the fair market price, an offer once made will seldom be withdrawn or reduced. The time element involved or a desire to save expense of litigation will sometimes make the buyer more eager. Therefore, take plenty of time to study the agreements; you may be repaid in more ways than one. Under certain circumstances, it may be advisable for you to defend an eminent-domain suit, but this decision requires expert legal advice.

These considerations add emphasis to the discussion of title work in Chapter 12. You can control your position on ROW's that come up after you acquire the tree farm, but you must investigate the provisions of those ROW's existing at the time of acquisition. Previous owners may have signed away many rights you want to keep. Do not forget that, as long as it is used, the ROW is not available for the growing of trees.

LEASES OF CULTIVABLE LAND

Presence of cultivable land including pastures on your tree farm offers possibilities of immediate, substantial cash income, and these possibilities should be analyzed carefully. Some cropland produces an annual cash rental three to five times the value of the timber that could be grown on the same land, and this money is available now and not 30 or more years later. Your first problem is to find out how much rent the cropland will produce.

The best way is to ask for bids from capable farmers of recognized financial ability. Chances are that they are already familiar with the productivity of your fields and that they are experts in making use of information available from the Soil Conservation Service. They are usually willing to bid on leasing them, since the cost of bidding is small, and you can solicit bids by means of a letter stating what you desire. The local bank can give you names and addresses of farmers who might be interested. If you get several bids on your land, you probably do not need any further information about prices. If you receive only one, you might want to check the price with the local county agent. Remember, however, that in most cases the best appraisal of value is a firm written offer of money, not an opinion from a third party.

Once you have an offer in hand, you should compare it with the income that might be obtained from planting the land in trees. A consulting forester with some knowledge of soil capabilities can help you here. He knows how much it will cost to establish the plantation, how much volume it will produce in a given period of time, how much this volume is likely to be worth when it is ready to harvest, and what annual taxes are now and what changes might be made in them after planting, and he is an expert in using compound-interest tables to bring all these future values down to the present for comparison. You should consult your tax adviser too, for income from land rent is usually ordinary income, whereas sale of timber produces a capital gain. Although this procedure sounds complicated and expensive, it is not so in practice, especially if the sums involved are small. In general, annual cash incomes are very effective in raising the value of a tree farm, since the money comes now, not later. Therefore, in most cases, you will decide to lease cultivable land.

Once you decide to do so, you should ask your lawyer to draw up or review the lease, since it is an important contract and may have a longer term than most. As always, all details should be covered in writing so that there will be few arguments later. The lawyer may be familiar with those provisions that vary with the kind of crop involved, but it is a good idea to compare notes with other lessors in your situation. Your bank can help you locate them also. Some leases should be for nothing less than cash in advance; others may call for some cash now and more after harvest; still others may be based on a share arrangement. Customary practices in the area usually establish a broad framework, and the experience of others will help you decide on the best plan.

RECREATION LEASES

Hunting and fishing leases often provide rentals that are relatively small, about equal to annual ad valorem taxes, but tree farming is a penny-pinching business and no annual cash income is unimportant. In addition to receiving rentals, you can often obtain help from lessees in preventing forest fires and trespass, since these two things are contrary to their interests also. As a general rule, good hunting and fishing are by-products of tree farm operation, and neither the game nor the lessees interfere with the main business.

You should consult your lawyer about these leases also; they are usually for several or many years and may make it difficult to sell the tree farm at some later date. You may also encounter a change in your public liability risk. The lessee is usually a club or association with many members, and a well-drawn written agreement will save many future arguments.

Our population is expanding rapidly, and demand for such leases seems to increase even more rapidly. There are more people with more spare time and more money to spend on leisure, and you should make some investigation of possibilities now. You might be able to negotiate a fortunate lease by spending a small sum to develop necessary facilities. Lakes for fishing and duck hunting and ranges for hunting doves and quail are obvious possibilities on small tree farms. Long-term, low-interest loans for some purposes are available through the Area Redevelopment Administration and other federal agencies, and you can get necessary information from the local office. This development may cause some loss of timber income, and your consulting forester can help you analyze this.

Hunting and fishing are just two aspects of recreation that produce income for the tree farmer; opportunities are limited only by the imagination of the owner. Developments of this sort are often highly specialized businesses requiring the active participation of the tree farm owner. I refer you to the published proceedings of the 1963 annual meeting of the Society of American Foresters; several papers on forest recreation were presented, and they may give you ideas you can apply. The call of the wild is still strong today, and none can predict where it will lead. The only sure thing is that recreation on tree farms will expand, and alert tree farmers should capitalize on these chances for profit.

GRAZING LEASES

Domestic animals have grazed the woods since early times, and, although research has demonstrated the superiority of improved pastures, many stockmen allow their animals to run on timbered land, at least during some seasons. Under certain conditions, this practice can be profitable for both cattlemen and tree farmers. The first grazing leases probably came about more to cure adverse possession than to produce revenue, but the shrinking land area available for grazing has made revenue more of a factor. In many cases, the rental in grazing leases is enough to pay annual ad valorem taxes.

Grazing is also beneficial because it reduces fire hazard. Hungry animals eat grass and shrubs that would otherwise serve as fuel. They may also eat young trees or trample them, but this can be prevented by proper management. Your consulting forester can help you determine how many animals your land will support without damage to timber. The number will vary with species of animal and type of timber.

The rental is ordinary income, but it is payable annually in cash and may be the only possible source of income during the early life of the stand. Although necessary fences restrict free movement of loggers and firefighters, logging operations are infrequent, fences keep down trespass, and lessees can be persuaded or required to help fight or prevent fire. Demand for grazing leases is not large at present, and I think advantages so far outweigh disadvantages that you should keep alert for the chance to negotiate them. As usual, leases should be prepared or reviewed by your lawyer.

MINERALS ON OR NEAR THE SURFACE

Chapter 6 covers minerals but does not include those substances not usually considered minerals in the legal sense. By this I mean sand, gravel, dirt, bauxite, limestone, or any other substance found near the surface and extracted by open-pit mining causing severe surface damage. Deposits of this kind may produce large profits for you. Locating them is a major problem. The top of such a deposit may be visible, or its presence may cause enough difference in appearance of the timber so that it will be discovered in tree farm inventory or operations. General information about minerals of your state is available from the state geological department. Specific information about your property must be obtained by drilling for sample cores, which requires expert assistance. A geologist can help you.

The commercial value of such a deposit depends upon many things. First, how much soil must be removed to get at it? Second, how far is the deposit from the point of use? Third, what are the quality and extent of the deposit? All of these are complicated problems beyond the scope of this book, but there are some general principles I can mention.

The importance of a location changes. For example, construction of a new road near you may make a gravel deposit suddenly valuable because distance to the point of use is reduced. The gravel may be used on the road itself, or the road may allow a shorter haul to other places. In addition, deposits nearer the market may be gradually exhausted, which would enhance the value of yours.

As a general rule, material of this kind is paid for per unit as it is removed, since it is difficult to determine the size of a deposit before operations begin. This makes the character of the mineral operator important. The average tree farmer does not have time to be present constantly to check how much is removed. Other tree farmers nearby usually have had some experience with the operators and will share their experience with you.

A written agreement between you and the operator is a must, and your lawyer should prepare or review it. He may be necessary also to determine whether you actually own the material, since definition of minerals is a tricky business. Your tax adviser can tell you about tax treatment of this income. Percentage depletion usually applies to these minerals, and the percentage varies with the kind of material. Your consulting forester can advise you about what effect such operations will have upon the surface. They will prohibit tree growing during actual operations and may cause a permanent reduction in, or elimination of, the productive capacity of the area. Some rehabilitation may be possible when deposits are exhausted, and the contract covering operations should provide for this if possible.

OTHER SPECIAL SITUATIONS

Stumps from virgin timber trees on some tree farms are impregnated with resin and are desirable because steam distillation can be used to recover chemicals they contain. Their presence is easily detected in a tree farm inventory. Sale of such stumps provides extra income and may assist in establishing the new crop of trees. They are usually sold by the ton or in a lump-sum sale; the first method is usually preferable, because most of a stump is underground and its weight is hard to estimate. The commercial value of stumps depends upon their size, weight per acre, and accessibility. Any consulting forester can tell you their value and give you information about the character and operating methods of buyers in your area. Although stump removal causes some damage to standing timber, experienced machine operators can keep damage to a negligible amount.

Certain species of trees supply valuable products from sap extracted by scraping away the bark or boring into the tree, and income from this source may fit into your program. Sweetgums produce storax, maples produce sugar, and certain pines produce resin. Rights to extract these products are usually leased to specialists. Payment is according to number of trees worked, depending on their size. Leases vary in length, but commonly run for three years. These cash incomes are important as always, but there are some disadvantages. The lease ordinarily prohibits any other disposal of timber during its term. The lessee's operations usually reduce the incidence of wildfires, but injury to the tree as part of the production process may increase mortality and damage if a fire occurs, decrease growth, and decrease the value of the tree when it is finally harvested. Metal objects are often driven into the tree in this work, and operators sometimes leave a few there, thereby decreasing the tree's value for such a product as sawtimber. You should discuss these factors at length with an experienced forester before acting.

A most delightful source of tree farm income is the sale of Christmas trees. They are things of beauty; they have a wonderful smell; they give great joy to the buyer and, especially, to his children; they may be highly profitable. Few businesses have such desirable characteristics. Nevertheless, this is a difficult business for the average tree farmer, unless he sells trees to an experienced dealer. The selling of trees to individual users or retail stores is a highly specialized art. Buyers will bid against each other for your sawtimber, but you must use persuasion and salesmanship to sell Christmas trees. The growing of a desirable tree often requires pruning and cultivation, for few trees in natural stands are full enough to please the housewife. Deliveries to market must be precisely timed. Dried-out trees are undesirable, and nothing is more useless than a Christmas tree on December 26. If you plan to grow and sell more trees than are necessary for your family and friends, you should discuss the project with someone who has been or is active in the business.

6
Minerals

Minerals are important to every tree farmer, whether he owns those under his land or they have been reserved entirely or in part by previous owners. They may be extremely valuable or nearly worthless, and their value can change almost overnight. In recent years, the political structure of much of the world has undergone radical change. Because of the threat of nationalization, oppressive taxation, or political instability, mineral development in many overseas areas is no longer attractive. Consequently, exploitation of lower grade deposits in our own country is increasing. You must do all you can to enhance their value, provided these actions do not cause an equal amount of damage to other parts of the tree farm. Vast industries are based on minerals alone, and the subject is broad and complicated. The purpose here is to discuss some reasons why they are important to you and to point the way to further information. First, we need to know what we are talking about.

DEFINITION OF MINERALS

Webster's New Collegiate Dictionary (1959) defines a mineral as "any chemical element or compound occurring naturally as a product of inorganic processes," and we generally think of minerals as materials of which rocks of the earth's crust are made. Some believe a mineral is anything neither animal nor vegetable. For your purposes, the important thing is what the law determines a mineral to be, and this is a real wilderness.

The question of defining minerals has been before courts for years and probably will continue to be decided for many more years. When minerals are sold or leased, the seller, or grantor, usually knows what he is conveying, or thinks he does, but the buyer, or grantee, may not have the same idea.

45

These differences, often caused by careless wording or new mineralogical developments, may not be presented to the court for solution until many years later, when it is almost impossible to determine exactly what was intended. This causes trouble. Water, sand, and gravel are considered minerals by some authorities, and they would be so classified under the dictionary definition. Nevertheless, many courts have decided that these are not included under the term *minerals*, and most tree farmers do not consider them such. Other true minerals such as bauxite and limestone deposits do not come within the legal meaning of *minerals* under certain conditions. This is often the case with such minerals whose removal deprives the surface owner of the use of the land and actually destroys the surface. Other substances, formed by organic processes and thus excluded by the dictionary definition, are considered minerals by the courts. This is true of oil, gas, and coal, substances of primary importance to many tree farmers.

Other major difficulties are that the definition varies from state to state and that courts have decided the same question in different ways at different times. This complicated situation surrounding items of great actual or potential value is a strong argument for consulting an expert lawyer when you are buying a tree farm with a mineral reservation or reserving minerals in a sale. Exact wording of the reservation may not become vital until years later, when it is impossible to change it. Your position depends to a great extent on what fraction of the minerals you own, and the simplest situation exists when you own all of them.

FULL MINERAL OWNERSHIP

Theoretically, if you want to exploit the minerals yourself, all you have to do is dig a hole or drill a well to get at them, bring them to the surface, and sell them. This is seldom possible because of the difficulties involved and the large capital requirements; you will probably sell the right to exploit them under certain terms and conditions to someone who specializes in this. This is a whole new world, and you must understand some of its language.

Common Mineral Terms. The most common way to keep an interest in minerals and yet have them developed is to enter into a *mineral lease*. This is a written contract by which the mineral owner, or lessor, allows the developer, or lessee, to explore for and extract specified minerals from certain lands for a limited period of time and for certain considerations. It is often a lengthy printed form with blank spaces for the proper insertions, and you should read all of it carefully in spite of its length.

One blank space is for the legal description of the land, and it is often followed by the "Mother Hubbard" clause. This clause provides that the lease covers all land owned or claimed by the lessor that is adjacent to, although not included within the boundaries of, the land described in the lease. It is normally inserted to take care of those small areas around the boundaries that the lessor may own by adverse possession only. The clause can have serious consequences, and you should consult your lawyer about its meaning to you.

The period of time may be of any length, but it is commonly five to ten years, and it is known as the "primary term." Moreover, in the paragraph about the term, there usually is a provision extending the lease for as long after the primary term as minerals in commercial quantities are produced. In other words, the lessee has so many years to search for minerals and, if successful, may extract them as long as the operation is profitable. Many tree farmers want immediate development of minerals, on the assumption that production will make them more money than lease rentals, and this is generally true. On the other hand, if exploration reveals that minerals in commercial quantities are absent, the lessee immediately drops the lease, and rental income stops. Large sums have been paid to lessors in bonuses and rentals on minerals whose value has never been determined, and such payments may be sizable in your case. For the reasons discussed below, a short-term lease is better than a long-term one.

Consideration may take several forms. First is the *bonus*, a sum paid by the lessee to induce the lessor to sign the lease in the first place, and this usually gives the lessee the right to conduct development operations for a certain period, often one year. The amount of the bonus is important, since it may be all the lessor ever gets. Upon receiving the lease, the lessee is expected to begin operations within the specified period, and the lease is automatically terminated if he does not. The lessee can buy the privilege of delaying them for 12 months by paying the lessor in advance another form of consideration, commonly called "rental" or "delay rental." The lessee can repeat this process each year during the primary term of the lease. Failure to either begin operations or pay rental terminates the lease. Therefore, leases are one-way streets; the lessee binds the lessor for the term of the lease by paying rental but can cancel the lease merely by failing to pay it. You should not count on the rental until it actually arrives. Many lessees send a release to the lessor whenever a lease is terminated, and you should make sure that this procedure is followed. The lease is usually recorded in public records and is an encumbrance on the land until a release is recorded. Bonus and rental payments are usually expressed as a certain amount per acre of minerals, and, when you own a full mineral interest, the number of mineral acres and the surface area are equal.

As a general rule, the amount of the bonus is far more important than the amount of the rental. The bonus, paid when the lease is signed, is a sure thing; the rental is paid later, if the lessee wants to continue the lease, and is an unsure thing. In areas of interest, the bonus tends to be a much larger sum than rental. In addition, most mineral operations are speculative ventures, and new data obtained on an area may change opinions of lessees rapidly. In many cases, leases have been bought and dropped several times within a five-year period. The good trader puts himself in a position to negotiate for bonuses as often as possible.

Consideration may take a third form if everything goes well and production of minerals begins. This is known as "royalty" and is the portion of production that remains the property of the lessor and may be stated as a fraction of production or as a certain price per unit. Although the lessor may receive his payment in kind, the settlement is usually monetary. The term means the right to receive a share of minerals produced, free of all cost and risk, whether the operator makes money in their production or not, and, therefore, it usually represents a small fraction. The amount of the royalty is negotiated by the lessor and lessee and is set forth in the lease. As the development of certain minerals proceeds, the royalty fraction often becomes the same in all leases over a wide area; for example, most oil and gas leases provide three-sixteenths royalty. Local inquiry will determine what these usual fractions are, and customary practice is usually based on the distilled experience of years of such operations. Once production begins, royalty payments usually keep the lease in force after the primary term expires, and this explains why releases must be recorded. A title lawyer looks only at public records and has no way of knowing whether there was or is any production under the lease. As soon as production begins, you will usually encounter a slight delay in receiving royalty payments, to allow time to prepare a division order. This is an agreement prepared by the operator and states the ownership of each interested party. As soon as it is signed by everyone, checks come regularly as long as minerals are produced.

The foregoing discussion of bonus, rental, and royalty applies primarily to oil and gas leases; there are slight changes in the case of "hard" minerals such as coal. The bonus is paid for the exploration period; it is based on the area involved and is to compensate you for damages that might arise from prospecting and for the inconvenience of having your land tied up. The usual exploration period is one year. If the lessee then exercises his option to lease, there are annual payments in the form of advance-production royalties, which are deducted from future royalties. Royalties are paid on the basis of a percentage of the price of the product or so much for each ton or cubic yard extracted.

Royalties for different minerals vary widely, and you should investigate the prevailing prices paid in the nearest area where a particular mineral is being mined. Royalties of most common minerals range from 25¢ to 75¢ per cubic yard. (Most mineral handbooks contain conversion factors from cubic yard to ton and vice versa.) One acre of a deposit 1 ft. thick contains 1613 cubic yards. With a deposit 10 ft. thick and a royalty of 50¢ per cubic yard, a landowner would receive $8065 for each acre mined.

Your lease should require that you are to be notified well before mining operations begin. Such notice should give you time to sell the timber on the area without great sacrifice; six months is enough, a year is better.

Surface Damages Due to Operations Under the Lease. The process of searching for and extracting minerals almost always causes some damage to the surface and, usually, removal or destruction of whatever timber is on it. By-products of the production process may have the same effect. For example, oil wells often yield salt water along with oil, and salt water usually kills trees if it is allowed to flow on the land. Although lessees make every effort to dispose of harmful by-products, they are not always successful. The actual wording of the lease governs what payment must be made for this damage. Most lessees feel, and the usual printed form provides, that the consideration under the lease is sufficient to cover all damages caused in a normal operation, and they do not believe they should be required to pay additional sums except in case of extraordinary or unnecessary damage. Nevertheless, to avoid nuisance lawsuits and to build goodwill among surface owners, many lessees customarily pay whatever damages are reasonable for the disturbance caused by their operations.

When damage occurs, you may have an argument, or even a lawsuit, over whether it was part of normal operations or due to negligence. The best way to eliminate uncertainty is to write into the lease a provision requiring the lessee to pay for *all* damage done to the surface.

Exploration Permits. A lease normally permits whatever activities are necessary to explore for minerals, and the lease should provide for payment of damages caused by exploring. You might be asked for an exploration permit even though your minerals are not leased, and this would bring other factors into play. Exploration may reveal mineral possibilities that will cause a flurry of leasing activity and development, or it may show such negative results that all interest is killed. Only a mineral expert is qualified to advise you on the probable results of exploration, but many tree farmers think that fees should be charged for such exploration. The fee may be a flat rate per acre, or it may be based on the amount of damage caused to the surface plus the cost of appraising it. In rare cases, tree farmers may grant exploration permits in exchange for data gathered by the exploration party, but, since most lessees are reluctant to make this trade, and since the data may be meaningless to a layman, payment in money is usually preferable.

Customary Buying Practices. The possibility of acquiring quick riches from minerals brings a host of buyers to areas of interest. They are usually self-employed; some are outright speculators; many represent substantial independents or major companies. All are aggressive, hardworking people who want to buy leases, minerals, royalty, or any other mineral interest available. Pressure is on, and there is no time to find out whether the seller owns what he is selling. Most buyers are familiar with title work but not expert at it. Therefore, they pay for what they buy with drafts of a 10-day or longer time limit. This period allows lawyers working with them to examine titles; if the titles prove faulty, buyers do not honor their drafts. Occasionally, a buyer will use this practice when he thinks he knows where he can sell a lease but has no specific order for it. He will pay for the lease with a draft and try to sell it before the draft must be honored. Drafts not honored always make people angry, and reputable buyers do everything possible to avoid this practice.

Drafts are for the buyer's convenience, not the seller's, and you can avoid difficulty here by insisting on payment in cash. This may discourage some buyers, but a serious one will usually agree to this plan. Insistence on cash is especially important at times of intense interest, such as when a well is being drilled and is near the zone of expected production. Mineral values may fluctuate by hundreds or thousands of dollars per acre in several hours, and, by accepting a draft, you have given a free option on your minerals without any assurance that it will be exercised.

The decision about cash or draft depends to a great extent upon the activity in the area and the buyers involved, and either method may be quite satisfactory. If you accept a draft, however, you should present it to your bank for collection with the lease attached to it. If the buyer honors the draft, his bank will give him the lease when it sends you the money. If you give the lease directly to the buyer, he may record it, sell it to an innocent third party, and also fail to honor the draft. Your minerals may then be under lease, and you may have great difficulty recovering money from the original buyer. This is an unlikely occurrence, but it can be precluded altogether by keeping the two documents attached until the draft is honored.

PARTIAL OR NO MINERAL OWNERSHIP

Mineral ownership can be separated from ownership of the surface, and the situation can be further complicated by separation of minerals and royalty. Separation may be permanent or temporary and may be the result of state mineral laws or agreement between parties. It is accomplished by what are commonly called "reservations."

Provisions of the Reservation. The foregoing section shows how important the language of the reservation is and how much you need a good lawyer to study and explain it. You must have the answers to many questions.

What minerals are reserved? Were they reserved in general terms or was the previous owner an expert mineral operator who used a detailed and all-inclusive reservation? The reservation may be quite specific, so far as oil and gas are concerned, but make no actual or implied mention of other minerals. In unusual cases, the wording of the reservation may cover sand, gravel, and other surface deposits.

What fraction has been reserved? It is common for each successive owner of a property to reserve part of the minerals, and some people are not as careful or as good at arithmetic as you are. For example, A sells to B and reserves half the minerals; B sells to C, reserving half of what he owns; C sells to D, reserving half of what he owns; D, thinking he owns one-quarter of the minerals, sells to E and reserves one-eighth, thereby separating the last fraction; and E gets nothing. Proper choice of words in mineral reservations is a technical legal problem, and any of these four may not have reserved what he intended to reserve. Was the full mineral interest reserved or perhaps only the royalty?

What is the length of the reservation? Is it permanent? Some state laws prohibit permanent reservations under certain conditions, and the variety of limited reservations that can be made by individuals is endless. Many tree farms were owned at one time by the Federal Land Bank, and it is common to find reservation of half the minerals for 25 years by the Land Bank.

What access rights, implied or specified, accompany the reserved minerals? As a general rule, the law allows some access to reserved minerals even though the reserving words do not mention access specifically, and you need to explore this facet. The owner of reserved minerals may be able to lease them under such conditions that the lessee will be required to pay only for unusual or unnecessary damage done in operations.

These questions point out the value of even a small fraction of minerals to the surface owner. A lessee must lease all minerals under a tract to develop it, since the owner of an unleased fraction may be entitled to his pro rata share of minerals extracted without paying some costs of extraction. Ownership of any portion of the minerals enables a surface owner to make sure that mineral leases require the lessee to pay for all damage to the surface and, therefore, to protect his trees. In addition, it may have a great effect on future values, since few investors are willing to build houses, service stations, or industrial plants on land where they have little control over mineral operations.

We have discussed this as if the owner sold the surface and reserved the minerals. Of course, while retaining the surface, he can convey the minerals. All of the above applies in either situation.

Exploration Permits. Mineral exploration is usually performed for major companies by specialists who do the work on a contractual or other basis. These specialists have an assigned work area and often contact the surface owners in the work area by telephone to ask permission to enter their lands for exploration. As a general rule, they have no knowledge of mineral ownership in the area, including whether the minerals are leased by their client. If the surface owner allows them to enter his land, all is well; they are then concerned only with claims for any damages they might cause.

Your trading position determines what response you should make to such a request, and it depends on the language of the mineral reservation. Under certain circumstances, you may have no trading position at all, and it is wise to investigate this with the help of your lawyer. If you cannot do this, you can quote a reasonable fee to the next man requesting a permit and see what happens. He might agree to pay it, since his operations may cause some damage for which he knows he is liable, and his agreement solves your problem.

TRADING IN LEASED MINERALS

When buying minerals under lease, you must know the terms of the lease, since the lessee has prior rights, and a title opinion usually gives this information. You also want to receive your pro rata share of rentals. Most leases provide that no change in mineral ownership shall be binding on the lessee until 30 days after he has been furnished by registered mail at his principal place of business with a certified copy of the recorded instrument as evidence of the change. Title examination may reveal the proper address, or you might be able to get it from the lessee's local office.

The lessee usually sends lease rentals to a bank designated by the lessor. The lessee needs proof that the money is received, and the lessor may move or die or be careless about indorsing checks or signing receipts. Using a bank to receive these payments is a satisfactory solution. You must notify the bank to send the money to you. Rentals are usually paid annually in advance, and you may be entitled to a pro rata share of the rental for the year in which you buy the minerals. This is something you will have to work out with the seller; since you are buying them subject to the lease, the lessee cannot help you.

AD VALOREM TAX ON MINERALS

When minerals are owned by the surface owner, they are not ordinarily assessed separately for ad valorem tax purposes, but this situation can, and often does, change when they are separated. Treatment by taxing authorities varies from county to county, and you should investigate this. Many states allow owners of separated minerals, at the time of separation, to buy documentary stamps in lieu of all future ad valorem taxes. They are usually sold in the county where the minerals are located and must be affixed to the deed. In theory at least, failure to buy and affix stamps means that minerals will be separately assessed and sold for taxes if the taxes are not paid. Tax sales require public notice, but notice is given in the county where the minerals are located, and an absentee owner may be unaware of an impending sale if he does not subscribe to the local paper. I have also found instances in which the minerals were still on the tax rolls because of a clerical error in the tax assessor's office even though stamps were bought and affixed. County officials are conscientious people, but they sometimes make mistakes, and looking after your business is your responsibility. Ad valorem taxes paid by mistake are seldom refunded.

MINERALS AND ADVERSE POSSESSION

We have seen that separated minerals still have some effect on the surface, and the reverse is also true. Minerals can be acquired by adverse possession under unusual circumstances. Suppose that adverse possession begins on a tree farm in 1930, before the minerals are separated. The record owner sells the minerals 10 years later and finally sells the tree farm 20 years later. Eventually, the adverse claimant makes his claim effective and acquires the surface. Chances are he acquires the minerals too, since his claim dates from a time when the minerals and surface were together. This is one more argument for using a good lawyer and for proper operation of the tree farm. I have said elsewhere that adverse possession is a difficult and sneaky problem, and you see that it may affect areas not normally considered in danger.

WHO CAN HELP YOU WITH MINERALS?

One obvious answer to this question is "a geologist." He specializes in minerals, and you can see that the field is so large that specialization is essential. Most principles used in selecting any professional adviser can be applied to selection of a geologist. Your state geologist can give you a list of those practicing in the state, and you can assay the relative competence of each individual by diligent inquiry around the business community.

General information about minerals of your area is available from the U.S. Geological Survey, which has offices in every state, and from your state geological department. The amount of this information is tremendous, and you will probably need some help in interpreting it. It makes interesting reading and tells you something about the wealth that may be yours.

One difficulty along your path here is that geologists are often primarily concerned with the development of minerals, whereas you are concerned with making money from them as part of your tree farm operation. The two objectives are not always in harmony. Your best adviser is a tree farmer who has extensive and successful experience in minerals management. These people are rare, and the value of their experience and knowledge is often unappreciated. You will do well to search for, and listen to, them. Tree farmers often sell their properties and retain a small fraction of the minerals, and repetition of this process will build a substantial mineral estate. Good advice is even more important here. Mineral ownership offers you a chance at extraordinary profits and deserves serious study.

7

Management Costs

The value of a tree farm as an investment depends on the net income it produces. It is difficult to raise gross income by increasing timber growth, because it takes 15 to 20 years or longer to grow a tree that can be sold. Since income tends to be fixed, the role of management costs is vital. You should study them thoroughly in order to predict them for your initial financial forecast, and you should appraise them at regular intervals to determine what reductions may be possible. As has been pointed out, a successful tree farmer is necessarily a penny pincher. The business is a long-term one, and a small cost has a large effect when repeated every year forever. Let us examine some of these management costs.

AD VALOREM TAXES

Ad valorem taxes are unavoidable, constantly increasing, and often the biggest item in annual cost. Therefore, they deserve your careful study, particularly since you might be saddled with an inequitable tax burden that was inherited from previous owners or that has become inequitable because of changes you have wrought.

Ad valorem taxes are primarily for the support of county government and are the main source of income for government at this level. The situation varies from one county to another, and even by districts within the county, so it is impossible for me to go into detail on your personal situation. The general method of taxation is usually established on a statewide basis, however, and there are many similarities among states. Let us take a close look at the procedure in Mississippi, since it will illustrate several important features.

The owner of property on January 1 is assessed with taxes for the coming year. He declares the property he owns to the tax assessor before April 1, using forms provided by the assessor. The assessor spends the next three months entering assessments on the land roll, making sure that every square foot of land in the county is on the roll and is assessed at its proper value. The owner sets a value on the property when he declares it, but the tax assessor can change this if he disagrees, and he wins any arguments on this subject for the moment. Shortly after July 1, the assessor presents the completed land roll to the board of supervisors, the county governing body elected by the citizens. The board inspects it carefully, ratifies it, and offers it for public inspection. On the first Monday in August, the board meets to hear the complaints of individuals who object to their assessments. After making whatever adjustments seem necessary, the board forwards the land roll to the State Tax Commission. The Commission reviews the assessments to see that they are in line with those of adjoining counties. Although it has no power to adjust individual assessments, it can order the board to raise or lower the level of assessments on one or more classes of property. After completing its task, the Commission returns the land roll to the county, where it is delivered to the tax collector. Taxes must be paid during the following January to avoid penalties, but they may be paid in late December if the taxpayer desires. If the owner fails to pay taxes, title to the property eventually passes to the county and, through it, to other individuals or the state.

The state constitution provides that all property be assessed in proportion to its value, and you can see what an administrative problem this is, since tree farm values vary so widely. Taxing authorities have solved it in a rough but workable way by dividing all land into either cultivable or uncultivable land and assigning a value to each class. For example, the county may assess all cultivable land at $24 per acre and all uncultivable land at $10 per acre. Improvements such as houses and barns are assessed separately and added to the land assessment. The tax rate, which varies slightly each year, is applied to the assessment, and the amount of annual taxes depends primarily on the assessment.

This system works surprisingly well, considering its obvious inequities. Nevertheless, as years go by, timberland assessments gradually approach the level of farmland assessments, whereas the market price of the former is often less than half the price of the latter. This means that tree farms bear more than their proportionate share of the tax burden. Although they are illegal and uncommon, there are sometimes discriminatory assessments against larger landowners who live outside the county. Assessments of many properties were made long ago and have never been changed. Houses that disappeared years ago and former cultivated fields that now contain a stand of trees 20 ft. tall are still on assessment rolls at their old values. The administrative burden on the tax assessor is heavy, and some errors are inevitable. A tree farmer who makes a thorough inspection of his assessment can often reduce his taxes considerably. Using qualified professional appraisers, one county recently completed a thorough reappraisal of all taxable property, and tree farm assessments returned to their fair level. No tree farmer expects preferential treatment, but all taxpayers are entitled to fair treatment. When assessments become outdated, you should encourage any movement in your county to hire qualified professional appraisers.

In Mississippi, early in this century, standing timber was taxed at higher and higher rates, and it was a race to see whether taxing authorities or sawmill operators would actually harvest the value of trees. This led to destructive cutting practices, idle land, and idle people. In an effort to place the tax burden on the tree farmer when he is most able to bear it, the legislature exempted timber from ad valorem tax in 1940 and substituted a severance tax due when the timber is cut. Other states have similar laws, some of which allow the tree farmer several options. Taxes of this kind must be paid by the tree farmer and included in his financial forecast. Severance or yield taxes and ad valorem taxes receive different treatment under federal tax laws. The first reduces the total consideration in a sale; the second is a deductible expense.

Your state tax commission and county tax assessor can supply any necessary information about ad valorem taxes on your property. You should study it. Taxes rise steadily as citizens demand increased services from government. Tree farmers pay a large part of the total tax burden but are generally not active in protesting proposed tax increases. Your county government will respond favorably to a protest based on facts and sound reasoning, and your familiarity with the tax structure will help you here as well as help you receive fair treatment under existing conditions. Remember that saving 6¢ per year on taxes increases net income by 6¢ and adds almost $1 to the value of your property.

FIRE PREVENTION AND SUPPRESSION

Most tree farmers rely on the state for fire protection, just as most urban residents rely on the city fire department. Fire-suppression activities are out of the question for the average tree farmer because of the heavy expense, and some larger tree farmers may benefit from a fresh appraisal of their suppression organizations. How much does fire suppression really cost, and what benefits are actually achieved? Cannot the job be done better and more cheaply by the state?

When firefighting organizations are operated by state forestry departments, they usually provide protection only in those counties where help has been requested by county government. Effective fire control is essential for successful tree farming, and every individual speaks with a loud voice in a situation like this. If your county has no protection, you should urge your elected officials to do what is necessary to provide it, and you should solicit the support of your fellow citizens for your efforts. If you already have fire protection, use the same methods to ensure that it is continued. Look up the telephone number of the dispatcher's fire tower in your area, and report all fires. It is a good idea to call the man in charge of the firefighting organization, tell him where your land is, and mention that you will appreciate all the help he can give you. He may well respond to your personal interest in him.

Fire prevention offers a possible field for individual action by tree farmers, regardless of the size of their property. One obvious thing you can do is construct firebreaks. This is expensive, and construction methods and resulting benefits must be carefully studied. Many people build a firebreak by making one or two passes with a farm tractor and a disc plow, on the theory that something is better than nothing. Since something costs money and may not be effective, it can be worse than nothing. Plowing two furrows 40 to 50 ft. apart and burning the strip between them are often necessary to afford any protection at all. Each protection situation demands a separate analysis, but you will probably find that construction of firebreaks is not economical except to protect plantations during early stages. This is particularly true when you use controlled burning as described below.

Another apparently effective tool is the offer of a sizable reward for information leading to conviction of anyone for setting a forest fire. The reward is usually offered by a state or local forestry association, but funds to pay rewards must be pledged by interested individuals. The chance that a reward will be paid is slight, and you can participate in a program of this sort at a small cost if the total of pledges to the reward fund is large enough for your pledge to amount to less than 10% of it. Although it is impossible to measure the effectiveness of such a program, I recommend that you participate if one exists in your area.

CONTROLLED BURNING

A relatively new forest-management tool is fire under control, and it promises financial benefits that may be greater than the biological ones. The practice is good business because it reduces risk. This applies particularly to plantations, and you should use this tool as soon as possible after planting. As trees approach 15 ft. in height and begin to form a solid canopy, there is a large accumulation of fuel on the ground, which can support a fire hot enough to cause a total loss. Removal of this serious threat early in the life of a plantation is almost mandatory. As trees grow taller and larger, they become less susceptible to destruction by ground fire, but periodic removal of accumulated fuel is still highly desirable. Wildfire always threatens disaster, and you should reduce your exposure whenever it is economically possible. As you will see later in this chapter, fire insurance on timber is expensive. Many investors make allowance for fire hazard by increasing the interest rate used in the financial forecast, and any decrease in fire hazard raises the value of the investment.

Controlled burning also reduces costs because it prevents establishment and development of some undesirable species. Hardwoods, for instance, frequently invade pine plantations and must be removed eventually. Periodic fires assist in this problem by preventing their initial establishment, thereby reducing or eliminating the cost of removal later. If allowed to grow unchecked, hardwoods may seize a significant portion of available growing space and thereby reduce the value of the final harvest, and this biological fact adds extra weight in favor of fire. The important thing to remember about use of fire for hardwood control is the role of interest. Money for small expenses now would grow to large sums if it were allowed to gather interest for 30 years, and you may not be able to justify use of fire for hardwood control alone.

The cost of such controlled fires is a deductible expense, and this feature is attractive. Fire may also be advantageous even when its cost must be capitalized, as is the case when it is used in establishing a plantation. A fire just before planting permits better spacing by providing better visibility, eliminates the possibility of wildfire for the first few months, and may be the cheapest method of hardwood control.

Of course, fire is a dangerous tool that should be used only by experienced people. It may consume what you are trying to protect and may even escape to adjacent properties. Before using it, you must notify the local firefighting organization and obtain a permit if necessary. Any fire at all is highly destructive in the management of hardwoods. Controlled burning may not have a place in the management of some other forest types, and it may be illegal under certain conditions. For these reasons, I recommend that you use it only after carefully planning with expert assistance.

BOUNDARY LINES

It is difficult to overestimate the importance of knowing exactly where your property is. Much of the United States was originally surveyed over 150 years ago, and dozens of times since, but boundary-line disputes still cause much friction and some violence. You can eliminate almost all of these disputes with a simple, inexpensive program.

First, you must locate the lines. This may be no task at all, since nearly every property has been surveyed at least once and the evidence may still be on the ground, although the work may have been done many years ago. The tree farm inventory may show the condition of the boundary lines. Your neighbors can often give you much information, since proper location of your lines is equally important to them, and this is particularly true if your neighbor is a large industrial tree farmer. A competent consulting forester or surveyor can check information obtained from your neighbors within reasonable limits of accuracy by pacing or some other inexpensive measurements on the ground. If such a check indicates that the lines are approximately correct, you should accept them. This saves the expense of a survey and a heated argument if the survey shows that the line should be moved a few feet onto the property your neighbor has been considering to be his. Moving your neighbor's line or his fence is often much harder than stealing his wife.

If a survey appears to be the only or best solution to your problem, be sure to select the surveyor carefully. The best arrangement is a joint survey with both parties choosing the surveyor and sharing expenses. If you cannot do this, select a surveyor whose competence is well known to your neighbors, and notify them of your plans. He should have commonsense and be able to explain what he is doing in a convincing manner. Many states require that surveyors be licensed, and any surveyor can tell you how to get a list of licensees. Many counties have an elected county surveyor, and the qualification for this job is all too often charm, not competence. Beware of incompetents; they cause problems where none exists.

Once the line is located, it must be established in some permanent fashion. The most common method of marking a line is to blaze trees on or near it, showing by a cut or hack in the bark the side on which the line passes. Some system must be used to show the relationship between the line and those trees on or near it. As a general rule of law, those trees whose trunks are entirely on your side of the line are yours, and those trees through whose trunks the line passes belong to you and your neighbors. You may eventually want to cut all your trees even if they help to establish line locations; you can usually blaze or hack other trees nearby if necessary. Marks of this kind are long lasting and easily visible to an experienced woodsman, but most tree farmers paint blazes and hacks a striking color for emphasis. White is very satisfactory because it is rare in nature; yellow, blue, silver, and red are also used. Since a gallon of paint is sufficient for one or two miles, paint costs much less than the labor of applying it, and you should use a good grade of outside house paint or paint made especially for boundary lines. Many tree farmers always use the same color, but this is not mandatory. The important thing is to get the necessary quality as cheaply as possible. Surplus paint is available from many sources, and your consulting forester can advise you about its suitability. Before applying paint, shave off loose outer bark so that the growth of the tree will not make it shed the paint rapidly.

Points at which lines turn should be marked. A short length of 1-in. iron pipe is satisfactory, since it is relatively permanent, cheap, and easy to transport. Wooden stakes that resist decay, rocks, concrete monuments, and other markers may be used, and you should select whatever is barely adequate for the job. Elaborate corner markers and aluminum tags nailed to a tree near each corner to show location of the corner are needless refinements.

Five to ten years later, the paint will be dim, and the line will be obscured by the growth of trees and shrubs. Nevertheless, corners, blazes, and hacks will be visible, and they can be renewed and repainted cheaply. They should be renewed in exactly the same place, and, if possible, the same trees should be blazed and painted; this procedure is important as legal evidence of possession. Reports on boundary-line maintenance should be dated, should state specifically that this method was followed, and should include detailed information about fences on or near the line.

Legal or other reasons may make it desirable to erect signs warning against trespass along your boundary lines. Many types of signs can be made, and permanence is important because the cost of erecting them is much larger than the cost of the signs. The manufacturer of state automobile license plates usually makes highly suitable metal signs the same size as license plates.

TRESPASS AND ADVERSE POSSESSION

The seriousness of trespass has decreased markedly with steady improvement of the national economy, but the main factor in reducing it is maintenance of boundary lines as described. A well-painted boundary line makes it almost impossible for a trespasser to plead ignorance, and elimination of timber theft almost pays for maintaining the lines. The only other necessity is to have some competent person make irregularly periodic inspections of the tract in such a manner that people of the community know he is doing so. Your neighbors can be a big help by informing you or your agent of unusual activities on your land, and they will be glad to do so if you discuss your need with them.

Adverse possession is a more insidious problem, because it often begins innocently as an accommodation to a neighbor. For example, he may desire to fence part of your land so that his cows will stay off the road or will be able to get to a convenient watering place, and as a friendly gesture you allow him to do so. As years pass, he may feel that he has acquired some right and begin to use your property as if it were his. If either property involved changes hands, the situation becomes more complicated, since the new owner has no knowledge of how the fence came about and tends to accept the situation as it is. Eventually, you may have a full-fledged dispute if you want the fence removed to prevent its damage by timber cutters, to provide free access for other purposes, or to meet the requirements of a future buyer. A fence enclosing any part of your tree farm must be covered by a lease or some other document showing clearly that it is there with your knowledge and permission and covering all other legal aspects. Requests for fencing privileges deserve thoughtful consideration by you and your lawyer, and unauthorized fences such as might be discovered by regular boundary-line maintenance should be acted upon quickly.

Possession has been the subject of many books and countless law-suits, and I can stress its importance to all tree farmers only by a greatly simplified description of what is involved. As referred to in this book, *possession* is the condition under which the record owner exercises his power over a tree farm to the exclusion of other persons, and *adverse possession* is the exercise of the same power over a tree farm by someone else against the record owner. Therefore, possession is possession; whether it is adverse depends on your viewpoint. It may be all the title that is necessary, and it adds great strength to any title. To be effective under the law, it must have certain characteristics.

Possession must be consistent with type of use. It is your only title to the money in your pocket, and you exert it by carrying the money with you. Since you cannot carry a tree farm, you must exert possession in other ways. One of the best is cutting and removing timber, but equally effective may be TSI work, planting, payment of taxes, fencing, painting of boundary lines, and so forth. It must be uninterrupted. There is no interruption when title passes from one person to another, so long as each carries out whatever acts constitute possession. Possession must be exerted against everyone without exception. For example, if you fence your boundary line to keep everybody out and then tell one of your neighbors that you are unsure about the boundary line and that he can disregard the fence, your possession is not legally effective. It must be open in the sense that the property is held without concealment or attempt at secrecy, its ownership is not covered up in the name of a third party, and there is no other attempt to withdraw it from sight. It must be held in such a manner that any person interested can ascertain by proper observation or inquiry who is actually in possession. It must be continued for a certain period of time as established by law.

The principle of possession exists to prevent difficulties, not to make wrongdoing easy, and it solves several problems associated with boundary lines. First, when continued long enough, it firmly establishes their location.

Surveying can be done in a very accurate manner, but the low values involved in tree farm surveying do not warrant the expense for extreme accuracy. Consequently, three surveyors may survey the same line and put it in three different places, each varying from the others by a few feet. Changing line locations every time a new surveyor comes along is expensive, and it is unnecessary if you establish your lines and then exert possession on everything inside them. Eventually, all the surveyors in the world will not be able to change them. Carry your acts of possession right up to the lines. Do not, for instance, leave a strip of uncut timber along the line just to be sure that you avoid a squabble with your neighbor. Doing so is a clear indication that you are unsure of your ownership and acknowledges to a certain extent the interests of another. Do not erect a boundary fence anywhere except exactly on the line. Backing off for any reason means idle land, wasted financial resources, and lower profits, and may cause you eventually to lose the unused strip.

Second, it settles worries about minor variations in area. Suppose you own a tract that contains exactly 160 acres according to the original public land survey but is found by an accurate survey actually to contain 163 acres. Although the mistake may have been made in the original survey, chances are the extra three acres came from an unintentional encroachment many years ago on land belonging to your neighbors. It is often costly to find out where the encroachment is and then relocate all boundary lines in the area in an effort to correct it, and some adjoining owners may object to such a procedure. Your adverse possession against your neighbor makes this unnecessary.

Lawyers have much less trouble with the law regarding possession than they do with the facts. Facts are hard to prove, particularly when they concern events many years old, and a record of facts is something that you must accumulate as time passes. Prompt recording of all forest-management activities is a must for this reason alone. Your record of possession is also helped by proper assessment of your tree farm and by recording all timber deeds, oil leases, and mortgages in which you are grantor or mortgagor. Put these things on record as they happen, not years later, when it might appear that you are doing so in an effort to fight a case of potential adverse possession; get time on your side by making your record as you go.

Although adverse possession is a bugaboo for most tree farmers and can be a sticky problem, the good management and maximum use we have been talking about all along will almost eliminate any dangers from it. Fulfillment of legal requirements of possession is good forest management, since it involves full use of the tree farm. A tree farm inventory will reveal adverse possession in most cases, and you can act promptly to interrupt it. Good management will prevent its development in the future.

Please remember that the foregoing is simplified and general; the whole matter of possession is complicated and varies from state to state. Therefore, under no circumstances should you attempt operations in this field without consulting your lawyer. People are sensitive about location of lines and fences, and expert legal advice is mandatory.

FOREST MANAGEMENT

The necessity for good forest management has been hammered home to all tree farmers recently. As long as prices rise steadily, everybody is a good forest manager, but return to normal up-and-down movement of prices soon eliminates incompetents and places a premium on sound judgment. Its extreme importance is still incompletely realized. Tree farming is a long-term venture, and there is a tendency to think that there is no hurry and that a job unfinished today can be finished next month, next year, or during the next cutting cycle. There is also a tendency to think that growth will compensate for any mistakes and that, therefore, each individual decision is unimportant. The role of interest points out the danger of these fallacies. The long-term nature of the business *increases* the importance of each decision.

Execution of the first thinning illustrates this very well. Some foresters, particularly inexperienced ones, use the marking gun with timidity when boldness is required. They leave trees that should come out and rationalize this by saying that they will be able to correct errors of this type the next time around. Therefore, they postpone income that should be captured now, slow down the growth of the remaining stand, and reduce the value of later thinnings by reducing the size of trees that will be cut then. The combined effect of these errors is large enough to alter the profit picture substantially. This is only one of many possible examples, and it shows how expensive inexperience is.

Experienced forest managers can prevent losses often caused by premature or improper application of research findings. Forest-research scientists discover some exciting things, and these discoveries often receive great publicity. The scientists are always careful to state exactly what has been found and especially careful to point out the limitations of the experiment. These precautions are often overlooked by tree farmers, and the temptation to pioneer is strong if handsome profits seem possible. Failure is almost an inevitable consequence, and failure costs money. It is a good idea to go slow in applying new techniques and to seek expert advice even then.

Good forest management is mandatory, and you must provide for its cost in your financial forecast. Although you do all other work yourself, you will probably need help from an experienced adviser. If you elect the do-it-yourself method, estimate the cost of your time, your car, temporary labor, and other expenses, and put all of this into your forecast. Management costs money regardless of who does the work, and tree farms require good management.

ON-THE-GROUND INSPECTIONS OF THE PROPERTY

Inspections of the property at least once a year are essential to good management. The sooner you know about attempted adverse possession or damage from fire or insects, the better you can avoid losses from them. The best inspections are made by the tree farmer himself. An old saying puts it well: "The best fertilizer is footsteps of the gardener." Those who live far from their tracts or cannot visit them should make arrangements for inspection by others.

Our field inspection service is an example of what is needed. We first verify that the assessment is correct and then report when taxes are due. Next we walk the perimeter of the tract to observe condition of boundary lines. We also walk into the tract at intervals to examine timber conditions. Finally we make recommendations about action needed and report on stumpage prices in the area. On the back side of our report sheet we sketch a map of the tract and add the names of adjoining landowners as shown by the tax roll. In most cases these inspections are done annually, but timing is up to the landowner.

The form of our report is displayed on pages 67 to 68. The questions to be answered force the inspector to look at important matters. We write the answers in by hand; those shown are typical.

You can use this as a model to develop a form of your own, and your lawyer might suggest ways to improve it for your purposes. Each year's report should be filed with the forest management records discussed below.

FOREST-MANAGEMENT RECORDS

It is imperative to keep a record of all forest-management activities on your tree farm. This record may be quite simple, but it must be continuous and all-inclusive. You should require a written report from everyone performing work on your property and put this report where it can be inspected from time to time.

An effective way to keep these records is to use copies of the map that is part of a tree farm inventory. The first step in making such a map is to prepare a tracing, and you can obtain additional copies for a pittance. Select a symbol for each activity such as planting, thinning, harvest cutting, TSI work, boundary-line work, and so forth. Then select a color, or a combination of color and number, for each year, and plot each activity directly on the map as it is completed. Usually you will be able to keep a record of all activities on one map. Although this is a simple task, very few tree farmers follow even this procedure; you will profit by doing so.

Time passes faster than you think, and, without records, you cannot remember exactly when you performed certain work on your land. How can you appraise its results if you do not know how long ago it was completed? Records of boundary-line maintenance are especially useful. They not only serve as a guide for the next crew performing this work; they may be invaluable in fighting a case of adverse possession. Records are necessary. You may use different people in the management of your tree farm. People move away, die, or change employment, and you lose much valuable history if the only records were in their minds. Each new manager must gather complete new data if there are no records of the past, and this process is far more expensive than keeping records.

PLANTING AND CULTURAL WORK

Every crop must be established and cultivated to speed or increase the final harvest. Your job is to reduce the cost of these activities to a minimum by careful planning.

When you plant trees, you must analyze the spacing factor carefully. A slight increase in spacing greatly decreases the required number of trees per acre and, thereby, the cost of planting. Wider spacing reduces or eliminates the need for precommercial thinning, a process which is almost prohibitively expensive. Planting an extra number of trees per acre to allow for mortality is not the proper way to ensure an adequate stand, but billions of unnecessary and undesirable seedlings have been planted at enormous expense in accordance with just such muddled thinking. Do not do this. You can ensure an adequate stand only by means of careful planting practice and good luck with the weather. If weather causes a failure, plant again. If you reduce the initial cost of plantations, you will be able to replant those that fail and still make money.

The need for care makes it essential to select the proper men for planting, whether you use a contractor or your own men, and you must include cost of supervision in your calculations. Tree planters face the obvious temptation to discard some trees in stump holes and creeks to improve apparent production, but this risk is less important than proper care of all seedlings and rigid adherence to the spacing pattern.

Use of controlled burning to reduce undesirable species in the stand was discussed earlier, but many tree farms contain excellent stands of desirable reproduction that are completely overtopped by worthless trees. Such acres, although apparently productive, are almost totally idle and must be released.

JAMES M. VARDAMAN & CO., INC.

FOREST MANAGEMENT SPECIALISTS

STANDARD LIFE BUILDING
P.O. BOX 902
JACKSON, MISSISSIPPI 39205
601-354-3123

SAMPLE

FIELD INSPECTION REPORT

Inspector: *J. K. Graham* Date: *18 Oct. 1973*
Title: *Manager, Jackson Branch*

Owner of Tract: *James A. Ross*
Address: *Box 1500, Memphis, Tenn. 38102*
Legal Description: *NE¼, Section 31, Township 10 North, Range 3 East,*
Johnson County, Mississippi.

1. **Assessment:** Description correct? *Yes* Why not? _____
   No. of uncultivable A. *160* Assessment/A. *$ 8.00* Correct? *yes*
   No. of cultivable A. *none* Assessment/A. _____ Correct? *yes*
   Assessment of improvements *none* What are they? _____ Correct? *yes*
   If there are errors, what action is needed? _____

2. **Ad Valorem Taxes:** When is next payment due? *January, 1974. Penalties begin Feb. 1*
   If it is not made, when will land be sold for taxes? *mid-September, 1974*
   If sale is made on this date, when will title mature in new owner? *September 1976*

3. **Perimeter Inspection:** How are boundaries marked? Show on map on reverse.
   What additional work is needed? *Boundary lines should be brushed out, blazed,*
   and painted at conspicuous color. Total length is 2 miles. Estimated cost? *$ 300.*

   Who are adjoining landowners? Show on map on reverse.
   What evidence of trespass is there? Show on map and describe here: *none*
   _____
   What action is needed? _____
   What evidence of adverse possession is there? Show on map and describe here:
   Fences as shown. They are not plotted to scale. Total encroachment is less than 1/10 A.
   What action is needed? *Ask owners to remove them or sign lease.*

4. **Timber Conditions:**
   What kind of timber is predominant? Pine *65* % Hardwood *35* %
   Sawtimber: Large (16" & up DBH) _____ Small(10"- 16" DBH) *✓*
   Small sawtimber and pulpwood _____ Pulpwood and young growth _____

   Is a sale desirable now? *yes* What kind? *Sawtimber*
   How much in volume? *200-300 MBF* In $ *12,000 - 18,000* Why? *Growth rate of*
   these trees has dropped below 6% annually. Market prices are very strong.

   Is TSI work needed? *yes* Planting? *No* Controlled burning? *No*
   Where are they needed? Show on map and describe here: *After sawtimber sale, to*
   remove cull trees. Entire tract should be treated.
   What is estimated cost? *$ 1,000*

   Is there evidence of damage by insects? *No* Fire? *No* Disease? *No* Wind? *No*
   Animals? *No* Ice? *No* What corrective action is needed? _____

   What are approximate stumpage prices in the area?
   Sawtimber per MBF, Doyle scale: Pine *$80 -120* Hardwood *$ 20-40*
   Pulpwood per cord: Pine *$5* Soft Hardwood *$3* Hard Hardwood *$ 1.50*

Figure 1

Methods of release are varied, and some are expensive. You must select one that is barely adequate to do the job. Everyone has seen beautiful stands of pure pine reproduction from which every hardwood sprig was painstakingly removed. The financial analyst cringes at the sight; the high cost of producing this eyewash so early in the crop virtually eliminates the possibility of future profit. You must seek expert advice before attempting cultural work of this kind.

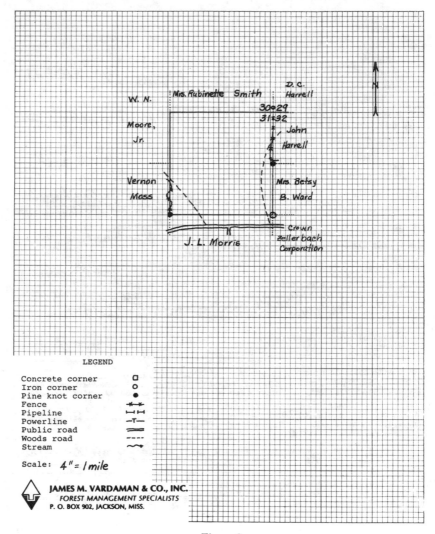

Figure 2

INSURANCE

Tree farmers, like all other businessmen, are subject to risks. Some of these are so small, and some so very large, that they must be assumed; most, however, can be transferred. Let us consider how to do this.

The most common method is to buy insurance. Hundreds of companies offer insurance applicable to tree farms, and they are usually represented by independent agents. Choosing the right company and agent is important. In judging both, you must consider experience, reputation, loss-adjustment records, and cost. You will find a certain degree of specialization among agents, since the field is so broad, and it is wise to consult one familiar with tree farm operation, even if he lives in another part of the state. Since insurance regulations vary widely from state to state, you must consult an agent in the one where your tree farm is located.

The average tree farmer needs only some form of public liability insurance, since his exposure to loss is slight. The cost is very low, and you may already have adequate coverage under policies you bought for other purposes. One great value in all insurance lies in the agreement by the insurer to defend you in all lawsuits. Anyone can sue you, even if the allegations in the suit are completely groundless, and you must either spend money to defend yourself or allow the plaintiff to obtain a default judgment against you. The cost of such a defense is many times the cost of insurance.

Workmen's compensation insurance laws usually vary by states and constitute a real wilderness. The normal tree farmer is not subject to them, but you should investigate this matter, since certain activities may make them apply to you. Your agent can tell you.

Insurance to cover loss on standing timber from fire and lightning is available from some companies and may cover all trees larger than 1 ft. tall. Rates are calculated by increasing or decreasing a base rate because of certain unfavorable or favorable conditions. Increases are caused by naval-stores operations, lack of fire-control organizations, recreational use of land, presence of unmerchantable trees, heavy underbrush, steep terrain, and other recognizable hazards. Decreases are caused by state fire protection, large tree size, fire resistance of species, heavy forest, light underbrush, and so on. This calculation is a complicated one and must be made for each tree farm. Policies must contain the exact legal description of the tree farm and the species of trees covered. You cannot insure timber so inaccessible that it cannot be logged by usual methods at ordinary profit margins. Reproduction is usually valued at a flat figure per acre and cannot be insured if it is located under a fully stocked stand of merchantable timber. The company often requires that the amount of insurance equal 90% of the value of the timber, and for each loss it is liable only for the difference in value before the fire and salvage value immediately after the fire. In order to reduce the number of small claims, policies usually contain a deductible clause that eliminates liability for losses less than $200 or 1% of the face amount of the policy. A dry-season charge allows the insurer to earn most of the annual premium during the hazardous months of the year. Your insurance agent can give you all the details you may require. The base rate is usually over 2% annually, and it is quite common for the total rate to be 3%. Therefore, the cost of this insurance can easily be a third of annual income; this explains why it is seldom used.

Most tree farmers provide their own fire insurance by increasing the interest rate used in the financial forecast, constructing firebreaks during critical periods, offering rewards for firebugs, or gambling that they will not have a fire. The increasing effectiveness of state fire-control organizations has greatly reduced risks, and a continuation of this trend is essential to successful tree farming. Fire insurance can also be carried under special programs by associations or loss-sharing arrangements among individuals, and there has been some study of expanding the federal crop-insurance program to include tree farms. Rates have already been well established by private industry, however, and they cannot be made much more attractive without public subsidies of some type.

SUMMARY

The value of tree farms as investments fluctuates with their net return, and reduction of management costs, wherever possible, increases net return. At a 6% rate, an increase of $1 in net annual income raises the value by almost $17. Frugality must be a day-to-day companion in tree farming and is a prerequisite for success.

8

Inventory

The first step in any business is to find out what it owns, what it has to sell. Many stock-market investors can recite from memory the exact number of shares they hold; some tree farmers have only a vague idea of what they own, and often they make no effort to find out. They neglect this step because tree farm inventories are sometimes called timber estimates, and many tree farmers have little or no merchantable timber. Actually, a tree farm inventory covers many things other than merchantable timber and is essential to success; let us discuss what it should include.

TIMBER INVENTORY

First in importance is the timber inventory. You may take it by measuring every tree on the property, but usually you measure an adequate sample of the trees, and you need to know something about sampling procedure. Sampling is based on the law of probability, and this law works well if you give it a chance to operate. If your tree farm contains 40 acres—15 open and 25 timbered—there is an obvious danger in basing your inventory on the measurement of only one sample. If the sample falls in the open area, you will show no timber at all, and you will overestimate the timber volume if the sample falls in the timbered area. If the sampling procedure calls for taking 40 or 400 samples evenly distributed over the tract, you can see that the law of probability has a chance to work and will produce an accurate answer. Inventories are described by the percentage of total area measured, such as a 10% or a 20% inventory. If you take a 20% inventory of a 40-acre tract, you measure every tree on 8 acres.

The sample must be properly distributed over the tract, and this is usually done by measuring strips or plots in a grid pattern. Statistical theory shows that samples taken at random may be used under certain conditions, but this practice is for experts. The important thing for you is to be sure that the sampling percentage is large enough to give you a reliable answer. The size of the sample depends on the total area, the purpose for which the inventory is made, and the species, size, distribution, and value of the timber. Generally speaking, the comprehensive inventory we are talking about here demands a 20% sample for areas under 300 acres and a 10% sample for larger areas. Smaller samples may be used when timber volumes alone are required. The report must show the sample percentage and how the sample was obtained. All other specifications must be covered in great detail.

Once you select the sampling procedure, you must decide how much information to gather. A simple statement that the tract contains a certain amount of pulpwood and sawtimber is suitable only for elementary use. The inventory must show volumes by products, species or groups of species, and DBH classes. If the tract contains more than several hundred acres, it should be divided into smaller blocks for the purpose of inventory. You need to know where the timber is in order to plan future operations. The report should clearly describe the units of volume measurement.

It is most important to know the number of trees by DBH classes and species. The size of a tree has a decided effect on its value, and you might need this information to take advantage of fleeting market opportunities. It is essential for growth studies and is valuable in comparing the present inventory with others that might be taken later. The inventory should give this information for all trees 5 in. DBH and up (generally considered merchantable timber), and it might be helpful to have data on trees 3 to 5 in. DBH, since they will soon become merchantable. General comments on trees less than 3 in. DBH are sufficient, since these trees will not become merchantable for 10 years or more.

VOLUME MEASURES

Pulpwood Volume Measures. The standard measure is the cord, a stack of wood 4 ft. × 4 ft. × 8 ft. containing a gross volume of 128 cubic feet. In many areas the unit is the common measure. A unit is usually 4 ft. × 8 ft. × 5 ft. 3 in., containing a gross volume of 168 cubic feet, or 1.3125 cords, but its dimensions may vary from place to place. This is an arbitrary measure chosen for sound reasons and precisely related to the standard cord. You must be sure whether you are talking about cords or units and how big a unit is. Pulpwood purchases by weight are common, and weight per cord varies widely with species and geography. Purchasing mills or men who harvest pulpwood can give you these weights for your area, and they can be checked against U.S. Forest Service figures. In the near future, pulpwood may be bought in the form of chips; you need expert help to convert tree volumes into pounds of chips.

Sawtimber Volume Measures. For over 100 years, we have wrestled with the problem of measuring logs (and trees) to determine how much lumber might be manufactured from them. This is a complex problem, since a round log must be made into square or rectangular pieces and the pieces must be separated from each other. Using either formula or diagram, we have developed three systems, or rules, for doing this that are in common use now, namely, the Doyle rule, Scribner rule, and International rule. In addition, logs are often measured by measuring the lumber produced from them, and this is called "lumber scale." Using these four systems in measuring, or scaling, the same pile of logs usually gives four different answers; if the logs are small, they may scale 1.00 MBF (thousand board feet) Doyle scale, 1.25 MBF Scribner scale, 1.35 MBF International scale, and 1.45 MBF lumber scale. To make things even more complicated, the relationship between volumes shown by these systems varies with log size. In certain cases, Doyle scale volume is less than lumber scale volume for the same log. The rule used to measure logs has no effect on their value, for this depends on what products they will produce. It does affect calculation of the selling price, and you must specify what rule is to be used in your timber sales.

Other problems arise in actually measuring or scaling logs. This usually is done by measuring the length and the average diameter inside the bark at the small end, determining the volume by the rule, and adjusting it for whatever defects the log contains. Scaling of logs, however, is not an exact science, and you may find wide variation among scalers. Some measure one or both bark thicknesses at the small end; some measure diameter at other points on the log; some have great difficulty in determining the average diameter when the log is relatively flat instead of round; some raise the measured diameter an inch or two to adjust for extra long logs; all have difficulty in judging the extent of defect and the deduction that should be made for it. Accurate scaling of logs requires skill and experience.

Variations in log rules and scaling practices can have a great effect on the amount of money you receive for your timber. I recommend that you seek expert advice until you become thoroughly familiar with this subject.

MAPS

The report should include a map on a convenient scale such as 2 or 4 in. to the mile. The map should show everything that might be useful to the forest manager and should be in such a form that it will fit readily into an ordinary file folder and be easily taken into the woods. A satisfactory solution is to prepare it on standard 8½ in. × 11 in. sheets, even if several sheets are required to show the entire property. Maps are made to be used and *will* be used if the design is convenient. Although the map shows the location of the property, the report must contain the correct legal description. The report may include aerial photographs of the property, since these are available through the U.S. Department of Agriculture at modest cost. You can find out how to get them from the local Agricultural Stabilization and Conservation Service (ASCS) office. An excellent scale for all purposes is 1/20,000 (you may prefer 1/16,000), and you should not order larger sizes unless you want a wall display. The photographs can provide stereoscopic views cheaply; you will probably need expert help in ordering the proper ones and the proper equipment.

GROWTH STUDY

Another important part of the tree farm inventory is the growth study. Accurate calculation of growth is a technical process requiring the training of a statistician and the patience of Job, but you do not need complicated procedures to get a figure that will serve your purposes. Your primary concern is the growth rate your trees will maintain for the next ten years or less. You can arrive at this within reasonably accurate limits by determining first how long it takes your average tree to grow 2 in. in diameter. Take the timber inventory by measuring samples on a grid pattern; you can gather growth data at the same time. Using an increment borer, take a core from enough trees of various diameters scattered over the tract, and count the number of rings in the outside inch. The taking of increment cores is hard work, and you can reduce the amount of it if, by using your judgment, you take cores only from trees that appear to be average for the tract. Avoid suppressed trees and large, limby ones growing all by themselves. A few cores will show whether there is a significant difference in the growth rate of the various species present and whether the exceptional species are important enough to warrant a separate growth prediction. In the South, for example, it is usually adequate to know how many years, on the average, it takes all pine trees on your property to grow 2 in. in diameter. Once you establish this average figure, you are ready for the next step.

First, make a list of the trees you now have, showing the number of trees in each DBH class, the volume per tree in each class, and the total volume of each class. Next, assume that every tree on the property grows exactly 2 in. in diameter, and make a list showing the number of trees by DBH classes after this growth has occurred. Now, insert the volume per tree for each DBH class, the same figures you used in the first list. Then, calculate the volumes per DBH class as before, and add them. When you finish and put the lists side by side, you will have a table that looks like Table 8–1.

TABLE 8-1

DBH in Inches	Now			Later		
	Number of Trees	Volume in Board Feet Per Tree	Total	Number of Trees	Volume in Board Feet Per Tree	Total
6	5,000					
8	2,500			5,000		
10	1,000	24	24,000	2,500	24	60,000
12	700	52	36,400	1,000	52	52,000
14	400	96	38,400	700	96	67,200
16	250	150	37,500	400	150	60,000
18	100	210	21,000	250	210	52,500
20 & up	50	280	14,000	150	280	42,000
	10,000		171,300	10,000		333,700

Table 8-1 shows a 94.8% increase in volume during the period. If the increment cores mentioned above determine that the average length of this period is nine years, your timber is growing at a simple interest rate of 10.5%. You should use the compound-interest rate, however, and reference to an interest table shows it to be 7.7% for this example. Application of this rate to the present volume (or timber capital) of 171,300 board feet indicates that your annual growth rate is 12,200 board feet. Theoretically, this volume can be removed each year without dipping into timber capital. If the annual growth is not removed, it is added to capital and increases the following year's growth, since the 7.7% applies to whatever timber capital you have.

The table applies to sawtimber only, and there are no volume entries opposite 6 and 8 in., since these trees are below sawtimber size. They will grow to this size, however, and information about their number is essential. Note that 8 in. trees listed under "Now" moved into the 10 in. class under "Later." This method makes no allowance for trees that will die during the period. Although this loss is normally insignificant, mortality can be serious, and experience is necessary to appraise this factor. Most of you will select a qualified professional man to take your tree farm inventory, and these calculations are properly part of his job. My intention here is to give you a general idea of what he will do and a simple, workable system if you want to do it yourself.

SOIL ANALYSIS

An equally important part of the tree farm inventory is an appraisal of the soil with which you have to work. Farmers have recognized the great variance in productive capacity among farms for a long time, but tree farmers have not; they have generally assumed that one acre is about as good for growing trees as another. Nothing is farther from the truth. Some acres will grow twice as much as others, and immediate recognition of this offers you excellent opportunities. You can use this lack of general appreciation either to dispose of unproductive areas you own or to acquire land of high productive capacity at bargain prices.

Trouble comes in finding someone competent to make a soil analysis. The study of the effect of soils on tree growth is relatively new in forestry, and, although many foresters have some knowledge of the subject, few have applied it extensively. Your consulting forester can do this work or will refer you to a specialist if he does not feel qualified. Many industrial tree farmers use soil analysis in tree farm appraisals, and you might be able to obtain helpful information from their foresters near you.

If you cannot obtain any technical help, you can certainly use your powers of observation. The most common measure of the productivity of forest soils is what we call "site index." This is the total height in feet to which the main trees in a timber stand will grow in 50 years; for example, an acre with a site index of 90 will grow a stand of trees 90 ft tall in 50 years. Other measures of site quality are in use; terms and ages may vary with species and geography, but the principle is the same. Therefore, you can get some idea of the productive capacity of the soil by noting the height and general condition of the bigger trees on it. Tall trees growing along stream bottoms point to excellent investments; runts fighting for life on a gravel deposit or the top of a ridge indicate trouble.

This does not work when there are no large trees at all; in such a case, you must use other methods. Soil scientists, by studying the mechanical and chemical properties of soil samples, can often predict accurately the timber-growing capacity of a soil; this is a complex process requiring expert help. In addition, the U.S. Soil Conservation Service, in cooperation with your state agricultural experiment station, is preparing soil surveys by counties for most of the nation. The reports are about 1-in. thick and contain much valuable information that can be understood by a layman. They include detailed descriptions of the soils of the county and a discussion of their suitability for crops, pasture, engineering applications, and wildlife food and cover, and the newer surveys have a section on their suitability for tree farms. This section enables you to predict fairly well the yield you can expect from each soil. Most of the report is made up of aerial photographs on which the soil boundaries are plotted. These reports are useful for many purposes and are available at no cost. You may find that the survey of your county has been published; if not, you will benefit from studying surveys of adjoining counties, especially if they contain the section on timber. The local office of the Soil Conservation Service can tell you which surveys are completed, and a consulting forester can help you interpret what these data mean to you in money. Annual production depends on soil productivity and governs return on investment.

NEED FOR DEVELOPMENT WORK

Your tree farm inventory should include a report on stand conditions, particularly those that require cultural treatment. Success in tree farming demands that every productive acre be busy, and you need information on those that are idle or nearly so. It is easy to gather this information while taking a sample on a grid pattern. Areas requiring planting, TSI work, controlled burning, and thinning should be shown on the map. Information on thinning areas is most useful, since this leads to immediate income and more rapid growth or concentration of growth on better trees. Areas damaged by beaver or other animals, insects, or diseases must be shown. It is also helpful to have comments about areas ready for harvest, but the timber inventory will assist in showing this.

Although the map will show the location and extent of these stand conditions, the comments of the man taking the inventory are needed to determine the cost of performing cultural treatments and the benefits that will be obtained from them. Financial forecasts of development work will be discussed in the next chapter; I want to point out now only the need for basic data.

MARKET ANALYSIS

The report should show how much various forest products found on the tract are worth on the present market. General comments on the prevailing price of stumpage in the county or state are not sufficient. Timber values vary widely with quality and accessibility, and you need specific information on the value of your timber to make a financial forecast.

Many tree farms contain some material that is unlikely ever to become merchantable and some material that is not merchantable now but may become so in the near future. An example of the latter is oak pulpwood. For many years it had few markets. There is now an active market for it near some paper mills, and the market is expanding. The market for hardwood pulpwood of all kinds is growing rapidly, and the price will increase with demand. The report must contain data on material on the tree farm that is likely to become merchantable or increase substantially in price in the near future.

OTHER CONSIDERATIONS

You need to know something about several other matters, and one of these is the prevalence of forest fires. Signs of repeated fires are easy to recognize on the ground, and their presence threatens trouble for the investor. Fire statistics for the county or state are far less valuable than a few comments about your tree farm in particular.

Another is forest type. There may be several on the tract, each presenting special problems and requiring special solutions to obtain satisfactory profits. For example, fire usually has a place in management of some pine species, but it is almost always a disaster in the management of bottomland hardwood.

It is useful to know who the adjoining landowners are. Although a complete report is not a part of the normal tree farm inventory, some information can be gathered on the ground. The U.S. Forest Service and large industrial tree farmers make excellent neighbors, since both help you with forest-fire and trespass problems, and the latter may be good customers when you want to sell timber or land. Neighbors of this kind usually keep their boundary lines painted and relieve you of this management expense.

Of great importance is any information that indicates the presence of adverse possession. It may be a serious problem, and I emphasize that the man taking the inventory must watch carefully for this condition and report in detail any evidence of its existence.

During inventory, a man must start at one boundary line and may touch others during the day, but he spends most of his time in the interior, where they are not visible. Nevertheless, any information about their condition is useful, for they must eventually be established, and this cost must be predicted. Equally important is information about area. Although a tree farm inventory is not as accurate as a survey, it will reveal any appreciable variation from the area shown by the legal description.

The map should show and the report should discuss ROW's found on the tree farm. Location, width, and type of use are important for two reasons. First, the location should coincide with that shown by title records; otherwise, you may have title difficulties. For many county and private roads, and even some state highways, ROW agreements do not appear on public records. Second, ROW's are ordinarily unavailable for growing timber. This wasted space may be a significant percentage of total area and reduce possible annual production of tree farm products. If such ROW's are still on the assessment, annual ad valorem taxes are higher in proportion to productive areas.

All preceding comments point to the vital importance of intelligent observation during inventory. For this reason, the best season for inventory is the period when trees and shrubs are bare. An experienced man can take an accurate inventory at any time, but he has an easier job when visibility is excellent. It is also obvious that inventories based on aerial-photograph interpretation with little ground work are unsatisfactory for most purposes; too many important items must be visually inspected, and a camera in an aircraft 2½ miles above ground is not good enough. Photographs are valuable, however, for reconnaissance and management.

You may hear a great deal about continuous forest inventory (CFI), a system requiring frequent and intensive measurement of sample plots with data usually tabulated for computer analysis; it is currently enjoying great popularity among some foresters. As a general rule, CFI is suitable only for areas over 10,000 acres, may not provide all the information an average tree farmer needs, and may be more expensive than conventional methods. A decision to adopt this system should be made only with great care and expert advice.

MANAGEMENT PLAN

To many foresters, a management plan is essential, and by management plan they mean a written document covering many years and describing in detail all aspects of tree farm operation, such as method of cutting, volumes to be removed from certain areas at certain times, cultural practices, and so forth. Plans of this kind may be desirable for properties owned by the U.S. Forest Service or large industrial tree farmers; they are usually not necessary for you and are premature as part of inventory. Inventory gathers data to determine what plans are possible.

Good management plans depend on many things in addition to forestry. Your tree farm may be only part of your assets, and, although you expect it to produce an adequate return, you want it to do so in such a manner that the return from all your investments fits your needs as well as possible. At times, you may want a regular annual income; in adversity, you may need immediate cash income; in prosperity, you may choose to invest funds from current income, in anticipation of future profits. Once you have a tree farm inventory, you can sit down with your advisers on forestry, accounting, and law and prepare a plan for next month, next year, or next decade. The more you tell them about your financial objectives, the better they are able to suggest a suitable plan. The plan should be simple and as flexible as possible, and you should review it at least annually to determine its suitability. Management plans are essential to successful tree farming, but they must be based on all aspects, not forestry alone.

As you carry out your plan on the ground, you will work great changes in the inventory, and you should keep it up to date with office calculations and entries that show the effect of each action. Under such a system, your inventory will serve as a sound guide for eight to ten years. Eventually, however, a calculated inventory will be subject to substantial error and must be checked by another, physical, inventory. You will want to compare the next inventory with the present one as adjusted by the calculations, and you must use the same standards in both to permit this. Insistence that all specifications be described in detail, as suggested earlier, makes this possible.

9

The Financial Forecast

Everyone who buys a tree farm makes big money. They are not making any more land, so all you have to do is pay the market price and hang on; inflation and development will take care of you. This seems to be the lesson of the past. In the 30 years I have been watching the scene, timberland prices have always seemed too high today, but have proved to be quite low when you looked back on them 10 years from today.

But if making money is so easy, why are there so many disgruntled tree farm investors who could have earned better returns on other investments? Why are there even some losers?

There have been both winners and losers because tree farming is like any other business. Success depends on attention to elementary business principles, one of the most important of which is preparation of financial forecasts. You can appreciate importance of the forecast by looking at some past and present mistakes.

FALLACIES OF COMMON APPRAISAL PRACTICE

A very common approach to buying a tree farm is to break it down into its components, assign a value to each, and add these values to determine how much to pay for the whole property. Components usually listed are merchantable timber, reproduction, bare land, and minerals. Although each component has a definite value if it can be acquired separately, they cannot be readily separated in a tree farm. They operate as a team, and their joint action is what means something to you.

The first weakness in this approach is that it is difficult to assign correct values to each component, especially since some of them do not even exist. There is nothing wrong with component-parts appraisals; the problem is poor valuation of components. It is easy to determine the market price of merchantable timber but hard to apply it. If this timber is present in very small volumes, less than a cord of pulpwood or less than 500 board feet of sawtimber per acre, you cannot assign the present market price to it, since you cannot harvest it even if you give it away. Nevertheless, this is often done. There is a market for minerals, but the tendency is to base mineral appraisals on professional opinions instead of cash offers. Although reproduction cannot be sold, it has a value, but this value again must be an opinion. It is usually arrived at by predicting the value at some future date and discounting this back to the present. The most nebulous component to be evaluated is bare land, devoid of minerals and merchantable timber and reproduction. It is almost impossible to find such an item, and it can be evaluated only by an educated guess. For your purposes, bare land has a value only as a base for growing trees, although it might be worth something as a speculation. You often hear the remark that any land ought to be worth so many dollars per acre. Such an attitude in tree farm appraisals and purchases invites disaster.

Second, market prices reflect the opinions of others, whereas you are interested only in yourself. The market price of a new Cadillac is about twice that of a new Chevrolet; is it worth that much more to you? Market prices of tree farms are strongly influenced by land-acquisition programs of large industrial concerns. When the programs begin, they exert a decided upward pressure on prices; once they end, prices drop rapidly to about the level where they offer a reasonable return on invested capital. This drop is hard to see, because prices of tree farms are not quoted daily, as are those of listed stocks, and because sellers strongly resist it. Ownership of a tree farm is evidence in itself that the owner is a man of some means and can weather a period of weak prices for some time. When large concerns stop buying, and thereby decrease demand, the first result is a slowdown in trading activity until it reaches a virtual standstill. Buyers are unwilling to pay high prices, since adequate returns appear unlikely; sellers are reluctant to give up their previous ideas of the value of their properties and hang on, hoping for a rise in the market. Finally, pressure on some sellers gets so strong that they decide to meet current market conditions, and the price is often a surprising percentage below the peak. This process may take several years, but, when trading is at a standstill, the usual explanation is a weak market where price reductions are a must. Proper appraisal practice will help you recognize such a situation and determine what a tree farm is worth to you.

EFFECT OF THE INTEREST COST

Let us demonstrate this by a detailed study of an actual transaction. A lumber company offers for sale a tree farm containing 400 acres and sets the initial offering price by the component-parts appraisal method, assigning per-acre values of $85 for sawtimber, $35 for pulpwood, $50 for reproduction, and $50 for bare land with a substantial mineral interest, for a total price of $88,000. Strenuous efforts by the company and its agents produce many sightseers, but no buyers, for 12 months. The company finally realizes that its tree farm must compete with investment opportunities available everywhere, so it solicits an offer from an experienced investor. He investigates before investing.

First, he takes an inventory of the property and gathers all data discussed in Chapter 8. He uses this to make a detailed schedule of income the property can be reasonably expected to produce. A careful and objective forecast indicates a total annual cash income of $7000. A careful and objective study of management costs shows a total of $1000 annually. Therefore, after payment of these, his annual cash flow is $6000.

Next, he assumes that he will borrow the entire purchase price and pay interest at the going rate of 8%. This is a proper assumption, even if borrowing is unnecessary, for his money can be put to many other uses. This is a useful device in appraising any investment. If it will not pay the going rate, it is probably a poor investment. To simplify his calculation, he assumes that all income will be received and all expenses paid on the last day of each year. He plans to use the cash flow first to pay interest on the unpaid balance of the loan and then to reduce the principal. Table 9–1 comprises schedules showing his financial status at the end of each year at three possible purchase prices.

He notices immediately the powerful effect of interest. Interest cost consumes so much of annual income that his loan balance drops very slowly during the early years. A difference of $2500 in price (about 4% of the above prices) has a great effect on the appeal of the investment as years pass. An annual cash income assumes great importance; failure to sell enough timber to meet all expenses means that additional investment is required, and this further increases interest cost. Increases in annual income, even small ones, have a pronounced effect, since all of the increase can be applied to the principal. At the $60,000 price, increasing annual cash flow by $500 would allow him to pay off his entire loan in the eighteenth year. The table also shows why so many new purchasers begin to cut timber as rapidly as possible: They are anxious to reduce capital investment and interest cost quickly.

Although the tree farm looks attractive for the long term, our investor, realizing that no one can predict the future and that he may have to dispose of it sooner than anticipated, investigates the short-term outlook also. He estimates that he can sell the tree farm, after operating it for five years, for $140 per acre ($56,000). Therefore, he makes a small profit at a price of $60,000, breaks even at $62,500, and loses money at $65,000. If he has to pay

TABLE 9-1

End of Year	$60,000 Price Principal Loan			$62,500 Price Principal Loan			$65,000 Price Principal Loan		
	Interest	Payment	Balance	Interest	Payment	Balance	Interest	Payment	Balance
			$60,000			$62,500			$65,000
1	$4,800	$1,200	58,800	$5,000	$1,000	61,500	$5,200	$ 800	64,200
2	4,704	1,296	57,504	4,920	1,080	60,420	5,136	864	63,336
3	4,600	1,400	56,104	4,834	1,166	59,254	5,067	933	62,403
4	4,488	1,512	54,592	4,740	1,260	57,994	4,992	1,008	61,395
5	4,367	1,633	52,959	4,640	1,360	56,634	4,912	1,088	60,307
6	4,237	1,763	51,196	4,531	1,469	55,165	4,825	1,175	59,132
7	4,096	1,904	49,292	4,413	1,587	53,578	4,731	1,269	57,863
8	3,943	2,057	47,235	4,286	1,714	51,864	4,629	1,371	56,492
9	3,779	2,221	45,014	4,149	1,851	50,013	4,519	1,481	55,011
10	3,601	2,399	42,615	4,001	1,999	48,014	4,401	1,599	53,412
11	3,409	2,591	40,024	3,841	2,159	45,855	4,273	1,727	51,685
12	3,202	2,798	37,226	3,668	2,332	43,523	4,135	1,865	49,820
13	2,978	3,022	34,202	3,482	2,518	41,005	3,986	2,014	47,806
14	2,736	3,264	30,938	3,280	2,720	38,285	3,824	2,176	45,630
15	2,475	3,525	27,413	3,063	2,937	35,348	3,650	2,350	43,280
16	2,193	3,807	23,606	2,828	3,172	32,176	3,462	2,538	40,742
17	1,888	4,112	19,494	2,574	3,426	28,750	3,259	2,741	38,001
18	1,560	4,440	15,054	2,300	3,700	25,050	3,040	2,960	35,041
19	1,204	4,796	10,258	2,004	3,996	21,054	2,803	3,197	31,844
20	821	5,179	5,079	1,684	4,316	16,738	2,548	3,452	28,392

a 5% commission on the sale five years from now, he loses money at all prices above $60,000. He decides that the proper price is $60,000, and he makes this offer.

The lumber company is shocked; as you remember, it offered the property at $88,000, establishing the price by the component-parts appraisal method. The table shows that purchase at this price is fool-hardy. Unfortunately, some parts of the property had been purchased by using this appraisal method, and this is one reason why the board of directors was disappointed in the yield from this part of the company's assets. Surprisingly, purchases based on this method are still common among people who ought to know better, and disappointment is inevitable. Nevertheless, looking forward and not backward, the lumber company makes a financial forecast of its own, compares the return it receives from the tree farm with what it might receive if the money offered were invested in manufacturing equipment or working capital, and accepts his offer.

You may argue with his method, since the timber income may continue forever, whereas his loan will be repaid in just over 20 years. At a price of $75,000, predicted income covers management costs and provides an 8% return on investment, and he can repay the principal by selling the tree farm. Our purpose, however, is not to debate; it is to study an experienced and successful capitalist at work on an actual transaction. He uses no magic formula; he merely considers each factor carefully and puts the entire calculation on paper so that results are easy to see. He offers to buy only at prices at which a profit appears likely; if his offer is rejected, he looks for other opportunities. We want to know how he thinks, for he plays a role of paramount importance to tree farming as a business and to industries that depend on forest products. We also want to know how he thinks because this gives us a clue to how his banker thinks. This man has ready access to the sources of tree farm loans, and we want to know his secret of success, since all tree farmers may need credit from time to time. His approach typifies that of wealthy individuals in this country, and you will do well to follow in his footsteps. When he makes a long-term commitment, he insists that income from the asset pay all carrying costs and also allow him to acquire an equity gradually. He may not use the approach of large industries, but we are concerned about you, not them.

WHAT INTEREST RATE SHOULD YOU USE?

How high is up? It is impossible to answer either question for you; I can only point out several factors that come into play. First, you are caught on the horns of a dilemma, the solution to which brings out the artistry, as opposed to science, of tree farm investment. If you select too low a rate, your appraised value will be greater than the asking price of nearly every property offered. Almost every trade will appear advantageous, but, of course, you will earn practically nothing on invested capital. On the other hand, if you select too high a rate, every offering will appear overpriced, and you will not make any trades at all.

Second, the rate depends on your assessment of the plus and minus factors discussed below and your ability to allow them to work for you. One of the most successful investors I know believes that the long-term rise in the dollar price of land is so substantial that almost any purchase is a good one over a 10-year or 20-year period. Some studies of the history of land prices indicate that he may be correct, and all of us know buyers who paid what appeared to be ridiculously high prices and then, within a short time, sold their purchases for a 100% profit. Nevertheless, this knowledge does not help if you pay such a price that interest and management costs force you into bankruptcy before long-term influences have time to work. Most successful investors give great attention to the near future and expect to receive at little or no present cost whatever profits the distant future may hold.

We used 8% interest in the calculation above, but the investor uses a somewhat higher rate, since he pays 8% interest on his loan and acquires an equity at the same time. He does not worry about figuring just what this rate is. He approaches the problem by asking, "How long will it take me to own this tree farm outright—10 years, 20 years?" He accepts a payout in about 20 years in our example, and this may be the proper length of time for you.

Experience indicates that the proper rate for you is the bank rate current at the time of acquisition plus or minus 0.5%, depending on your individual situation. In analyzing a tree farm investment offered to you, use this rate in a calculation similar to that of our investor, and see how long it takes you to own the property outright. Length of the period is a useful measure of the investment's appeal, but there are no definite rules for applying the label of "good" or "bad." Individual situations vary too widely, and effects of the plus and minus factors discussed in the following sections are too powerful. It appears, however, that tree farms are excellent investments if the period is 10 years, good if 20 years, and poor if 30 years. Many investors believe that the plus factors outweigh the minus factors by a wide margin; if this is true, an investment originally predicted to pay out in 20 years may do so in a much shorter period.

Selection of the interest rate you want to earn is equally important if you already own a tree farm. First, it enables you to determine whether the asset is performing as you desire. Using our investor's approach, appraise your tree farm, and decide what you would be willing to pay for it if it were offered to you. If this value is less than the price at which it can be sold, you might be wise to sell it and invest the proceeds elsewhere. Second, the rate has a big effect on whether you cut a tree today or allow it to grow for additional years. The timber marker must have such a guide in order to mark your stands for cutting. The general considerations above govern selection of the rate for this purpose. Some modifications may be necessary to reflect available opportunities for reinvestment of funds and how much of the sale proceeds must go for taxes.

PLUS FACTORS THAT CANNOT BE MEASURED

Our investor has owned and studied tree farms for years, and he knows they have substantial advantages that are hard to measure with a ruler or a dollar sign. These excite him once he finds a property that passes his financial tests. First, tree farm income and expenses receive unusual treatment under federal tax laws. Second, tree farms produce profits from sources other than trees, most of which cannot be anticipated. Third, he may have a chance to sell portions of the property for conversion to higher use, such as farms, ROW's, or pastures. Fourth, wood is a cheap and renewable natural resource (at present, the cheapest source of cellulose), which intrigues many scientists. New discoveries are frequent and will raise the value of his tree farm. Fifth, he cannot lose his entire investment, since land cannot be moved or destroyed and always has some residual value. An investor in an oil well buys an asset that will become worthless sooner or later.

Sixth, and most important of all, he knows that land prices in the United States have risen continually since the Pilgrims landed. They may go up and down from year to year, but almost any 10-year period shows a sizable increase. Studies of price trends in all parts of the country are available, and some show an average annual appreciation of 10%. It is reasonable to assume that this trend will continue. The pressure of people on land makes prices go up, and our population is expanding.

MINUS FACTORS THAT CANNOT BE MEASURED

He realizes that tree farm investment has some disadvantages. First, it is a long-term venture. Investors prize flexibility and liquidity; they want to react quickly to sudden changes in conditions. Transactions in listed stocks are almost instantaneous, and market prices are in the paper every morning. Sale of a tree farm may require several weeks or months, and the exact price is not determined until you find a buyer. Second, timber prices fluctuate, and, since interest and management costs tend to remain fixed, a severe drop in timber prices can hurt him. On the other hand, he knows that any drop will not occur suddenly, and he has enough financial resources to postpone sales for two or three years. Third, technological change and shifts in the rate and direction of our country's development may cause his property to fall in value. Fourth, over the long term, prices of some forest products have risen faster than those of competing materials, and this decreases the market for tree farm products. Nevertheless, he still has some freedom of action, and he believes that the effect of the last two factors will be so gradual that he will be able to react to them in time.

WHOLESALE VERSUS RETAIL PURCHASES

The actual transaction described above was many times larger than the figures in the table and might be considered a wholesale purchase. This explains both the great care used by the investor and his insistence upon adequate return on capital. His approach is sound in every respect, and you should follow his procedure even though your transactions are in what we might call the "retail market." This is the market familiar to most of us, where the normal transaction is less than $50,000. It is the land of dreams; romance and impulse have a real effect on market prices, and market experts are found on every street corner. It offers exceptional opportunities and exceptional risks. Component-parts appraisals are commonplace, and the use of this method by others may help you sell a tree farm. Nevertheless, you should use our investor's approach to govern your purchases and insist on an adequate return when you invest even a small amount. If the return depends more on hope than facts, keep your money; you can get a better deal later.

One half-truth believed by many foresters and tree farmers is that a big tree farm sells for a bigger price per acre than a little one. This is used as a reason for paying premiums to acquire adjacent lands. Nine times out of ten, the opposite is true. There are thousands of buyers for 40-acre tracts, hundreds of buyers for 4000-acre tracts, but very few buyers for 40,000-acre tracts. Competition decreases as the consideration increases. In addition, investors with large amounts of capital are shrewd and intelligent, know well the power of money, and use their great resources to make advantageous trades that people with less capital cannot make. This is why they have so much money in the first place.

Another half-truth frequently repeated is that it is worth a sizable premium if a tree farmer acquires a small tract completely or partially surrounded by his land. If he keeps his boundary lines painted or if the tract improves access to his land or if his neighbor is a constant source of trouble, there is at least something to this. As a general rule, however, a trading situation of this kind strongly favors the owner of the large tract; he is almost the only market for the small one and should be able to buy it on favorable terms if he is patient. You will do well to remember these two things when you are trading in the retail market. Trading in this market can be highly profitable, but it is several steps removed from pure tree farming and, therefore, beyond the scope of this book. I can only mention its existence to you and state that all principles covered herein will help you. Their application will establish the true value of tree farms you encounter and show how much of their market price depends on other considerations.

FINANCIAL FORECAST APPLIED TO CULTURAL WORK

We have discussed the financial forecast, with emphasis on the role of interest, only as it applies to the purchase of tree farms, but the procedure applies equally to other management activities such as planting, TSI work, and seed-tree cuts.

Millions of trees are planted yearly, and their spacing on the ground is sometimes decided by what appears to be confused thinking. The 6 ft. × 6 ft. spacing that was once the norm requires 1210 seedlings per acre. At a seedling cost of $11.00 per thousand and a planting cost of $39.00 per thousand, you invest $60.50 per acre. Let us assume that you borrow money at 8% to pay for planting. The first harvest cut might be made 25 years later, at which time interest will have multiplied this cost to $414.33 per acre. If you liquidate the stand at this age, you must cut over 41 cords at $10.00 just to break even. By expanding the spacing to 9 ft. × 9 ft., you plant only 538 trees per acre and reduce the initial cost to $26.90. In 25 years, interest will increase this cost to $184.22, so you can break even at less than 19 cords per acre. This calculation ignores the cost of the land on which to grow the trees and the costs of management and taxes; do not forget that interest runs on these too. Our purpose is to show that the decision about spacing is important. Some plant extra seedlings to allow for initial mortality; this is poor planning. Seedlings do not die in an alternate manner, and planting too thickly makes some mortality imperative. Unfavorable weather or fire often causes failure of the whole plantation, and two plantings at 9 ft. × 9 ft. cost less than one at 6 ft. × 6 ft.. Some argue that wide spacing allows too much open-grain wood at the center of the tree and decreases the value of the final crop. Biology enters the picture here, and there are many published research reports to help you decide whether this is important to you.

You should make the same forecast when considering TSI work. Once again, interest makes the size of the original investment significant. Spending $10 per acre to kill scrub oak trees overtopping young pines is one thing; spending $40 per acre to kill every undesirable bush and shrub in sight is entirely different.

Interest has a bearing on the practice of leaving seed trees to start the next crop. Seed from these trees is not free, regardless of arguments to the contrary. Leaving an adequate stand of trees delays the income that can be realized from their sale; you must figure interest on this sum, which is often more than $60 per acre. The trees may continue to grow, perhaps in excess of the interest rate, but some usually die because of great changes in the biological situation. They must be harvested to recover what growth they make, and the future price per MBF may be lower because of small volumes per acre. They must be harvested also to prevent interference with the new crop, and this commits you to a future cut when markets may not be favorable. Removal of seed trees may destroy the reproduction that came from them. In addition, spacing of natural reproduction cannot be controlled, and overcrowding is common. Choice of this method of reproducing your stands requires careful consideration.

Cultural practices needed on your tree farm depend mainly on biology, but financial considerations should govern which to undertake first. Some expenditures such as those for TSI work are deductible; others such as those for planting must be capitalized. This has an immediate effect on your taxable income. Benefits from some practices may be realized five years later; returns from others may not be obtained for 15 years. The need for some practices such as planting may be obvious, whereas that for others such as pruning or precommercial thinning may require theoretical calculations by an expert. If funds available for development are limited, you must channel them into the practices that produce the highest return.

Finally, you must recognize that many acres are inherently productive but are idle because of poor management in the past, whereas some acres are idle because they are inherently unproductive. Those in the first class can be put to work with varying success by means of cultural treatments; those in the second may justify no additional investment whatever. In an extreme case, a tree farmer would be better off to give away his unproductive acres, since they will not produce enough revenue to pay ad valorem taxes.

10

Accounting and Federal Tax Laws

Accounting and taxation are the least understood of the important aspects of tree farming, and you must have some grasp of them. Your understanding helps your tax adviser help you and makes it possible for you to appraise some investment opportunities without outside help. Please realize that the following discussion touches only the high points.

INCOME

Capitalized Expenditures and Depletion. When you buy a tree farm, you swap dollars for natural resources. You have not changed the amount of your assets; you have merely changed their form. One day the list of your assets shows so much cash; the next day the cash is gone, and the list shows an equivalent amount of tree farm property. Transactions of this kind are capital expenditures. If you buy a tree farm for $1000, you capitalize the expenditure of $1000.

The transaction is more complicated than swapping dollars for tree farms. In the trade, you actually acquire some trees, some land, and perhaps some minerals, and you must show the amount paid for each in order to establish your depletion basis. In making such a trade, you establish a depletion basis by spending cash, but you must determine it whether you buy a tree farm, inherit one, or receive one as a gift.

Depletion basis is important because it determines what taxable income results from operating the tree farm. Let us suppose that, in the $1000 transaction above, you spend $900 for timber and $100 for land and minerals. If you sell all the timber for $900 the day after your purchase, you merely swap for dollars again and make no profit. In other words, you deplete your timber account by $900; it has a balance of $900, so there is no taxable gain.

If you sell the timber for $1000, or if you have no depletion basis at all, you have a taxable gain—a short-term gain if you owned the asset less than 12 months, or a long-term gain if you owned the asset more than 12 months. Short-term gain is ordinary income and receives the same tax treatment as your salary. Long-term gain receives special tax treatment, is commonly called "capital gain," and is so important in tree farming that it will be discussed at length in Chapter 11.

Once you capitalize an expenditure and establish a depletion basis for an asset, you cannot reduce or deplete the asset until you dispose of it. Depletion basis can never be more than actual cost, but whether the basis exists in timber or in land means much to you. If it is in timber, you recover some part of it every time you make a timber sale. If it is in land, you cannot recover any of it until you sell the land. Since you may sell timber every few years and dispose of the land only upon death, you should assign as much as possible of the total depletion basis to timber. Nevertheless, it is common to find tree farmers who buy a tract of cutover timberland for $100 per acre and assign $80 to land and minerals and $20 to timber. This is a bad practice and does not reflect facts. When you think of land in tree farming, you think of nothing but soil without any trees at all, not even the smallest ones. When you buy a tree farm for $100 per acre, you certainly want more than bare soil (unless the soil is exceptionally productive), and you receive more than bare soil, because the land has some small trees on it. Although they may not be merchantable, you expect them to become merchantable soon or you would not buy the tract at the price. Therefore, whether you realize it or not, you pay some of your money for timber, and your records should show this. In this example, it is much nearer the facts to assign $80 to timber and $20 to land and minerals.

Let us return to the $1000 example discussed earlier. If you assign $900 to timber and sell all of it at some later time, you deplete your timber account by $900 and show whatever additional money you receive as a gain. Complete liquidations, however, are rare; partial cuts are much more common. If you follow tax regulations exactly, this makes things more complicated. You must determine how much timber of each kind you own at the time of the sale and allocate part of the depletion basis to each. In this case, let us assume that you have nine MBF of sawtimber and nothing else; therefore, your depletion basis is $100 per MBF. If you sell any part of this timber immediately after purchase and receive $100 per MBF, you have only depletion of $100 per MBF. Suppose, however, that the cut is delayed 10 years and that, during the delay, timber volume increases to 18 MBF. At this time your depletion basis is only $50 per MBF, since the total cannot be changed from the original $900. If you sell any part of this timber for $100 per MBF, you have $50 of depletion and $50 of capital gain if you meet the requirements of capital gain treatment. You can see that calculations of this kind can go on forever and may become burdensome for owners of small tree farms. The easiest solution for small operators is to discuss with their tax adviser whether they can deplete the entire timber account in early sales and show all future sales as pure capital gain.

Values assigned to land as a depletion basis usually include minerals, if any. Your tax adviser can tell you whether you should allocate a separate depletion basis to minerals. Your tree farm may contain fences, buildings, barns, and similar improvements that can be depreciated. At this point, notice the difference between depreciation and depletion; they are sometimes confused. Depreciation is reduction in the service capacity of a capital asset through use, obsolescence, or inadequacy. Depletion implies removal of a natural resource—a physical shrinkage or lessening of an estimated available quantity. All of this makes establishment of the depletion basis a matter on which you must consult a tax adviser. The do-it-yourself method can get you in trouble with the Internal Revenue Service (IRS). If you failed to establish your basis when you acquired your tree farm years ago, it is not too late. You can do so with the help of a consulting forester, and your tax adviser can tell you how.

Tax Treatment. The foregoing paragraphs cover nearly every aspect of the tax treatment of revenue from sale of timber, the main source of income in tree farming. Some part of every sale is depletion, and some part is capital gain (or loss, in unusual cases), if you have held the tree farm as long as 12 months.

Other revenues receive varying treatments. In general, sales of ROW's are like timber sales but may also mean some depletion of the land account. Rentals under leases for turpentine operations, grazing, hunting, recreation, and so forth are ordinary income. Payments for damages are treated the same as timber sales. Option money is ordinary income, unless the option is exercised and the option money applies on the purchase price. Rentals under long-term timber leases require close scrutiny so that they can be properly classified. These leases are so difficult to negotiate and administer that most lessees are interested only in tracts of 25,000 acres or more, thus, this problem may not concern you.

EXPENSES

So far we have talked only about expenditures that change the form of your assets. Most of your expenditures, however, are not changes of form; they reduce the size of your assets. These are expenses, many of which are deductible and can be charged against any of your income. Now let us move into this briar patch.

Some expenditures are deductible; some must be capitalized; and a few may be treated in either way. Those in the last category are called "carrying charges" and "development expenditures." Rules about which items are deductible and which must be capitalized are not always clearly defined; there is a gray area where an item is deductible at one time and must be capitalized at another. Much hinges on the intent of the taxpayer and on timing. Certain expenditures are capital expenditures if incurred in connection with land acquisition or planting and are deductible expenses if incurred several years after acquisition or in connection with TSI work as opposed to planting.

Which Items Must Be Capitalized?   Capital expenditures form depletion basis and include

1. Acquisition cost of land and timber
   a. Lump-sum purchase price
   b. Inventory, boundary survey, legal and recording fees, and commission, if connected with purchase
2. Direct costs incurred in connection with reforestation by planting
   a. Preparation of planting site, including girdling, hardwood and brush removal, or other treatments to afford good growing conditions
   b. Cost of seedlings
   c. Labor and equipment expenses including depreciation or rental of tree planters, tractors, and so on, involved in planting
   d. Thinning and improvement cutting of unmerchantable timber, if in connection with planting
3. Permanent improvements to property, such as road construction, fire-break construction, grading, and ditching, if of a permanent nature with no determinable life

I mentioned earlier that capital expenditures must be allocated between land and timber accounts, and this applies to the three categories listed above. Your knowledge of tree farming probably gives you a general idea of the accounts in which the expenditures should be capitalized, but you should consult your tax adviser. You usually want to deduct as much as possible so far as total costs are concerned; if an expenditure must be capitalized, you usually want to capitalize as much as possible in the timber account.

Which Items Are Deductible?   Let us talk about something more pleasant, namely, expenses that are deductible items for income-tax purposes. They are more pleasant only to the extent that some portion of them would otherwise have gone for taxes. Deductible items are generally operating expenses incurred during the years a timber crop is being produced and include

1. Expenses for timber stand improvement, thinning, and improvement cutting in immature stands, where the work is not done in connection with planting
2. Cost of tools of short life (less than one year)
3. Cost of materials and supplies
4. Incidental repairs
5. Cost of temporary roads
6. Management expenses (for instance, use of your car for inspection trips to the tree farm)
7. Professional fees paid, if not directly in connection with a sale
8. Inventories and boundary surveys, if not in connection with property acquisition, and if not directly in connection with a sale
9. Ad valorem taxes
10. Interest paid on borrowed money (within certain limits)
11. Insurance premiums
12. Depreciation of equipment used in operations (subject to the ordinary rules for capitalizing and depreciating such equipment used in tree farm operations as trucks, tractors, power saws—generally, any equipment with a life in excess of one year)
13. Depreciation of improvements with limited and determinable lives, such as roads, bridges, culverts, and fences—logging roads, as opposed to permanent roads, being generally depreciable by either of two methods:
    a. So much per year of an estimated useful life
    b. So much per unit volume of timber expected to be brought out over the roads
14. Maintenance of firebreaks
15. Protection expenses against fire, insects; disease, and so on

Which Items Are Optional? Certain expenditures just listed in the category of deductible expenses do not, from a technical income-tax standpoint, come under the classification of deductible expenses; instead, they are classified as carrying charges and development expenditures. The government allows you to either deduct or capitalize them. I have included these items in the category of deductible expenses because situations in which you wish to capitalize, rather than deduct, them are so rare as to be of no practical interest. Generally, carrying charges include such items as annual taxes or interest paid on a mortgage. How many times have you wanted to capitalize, rather than deduct, taxes and interest? Development expenditures are those incurred in caring for timber stands in the development stage, while they are not yet productive of income, such as

1. Thinning, girdling, poisoning, pruning, and improvement cutting
2. Protection against fire, insects, and diseases, including labor, materials, and tools used in maintenance of firebreaks, and contributions to fire-protection associations

How Do You Treat Government Subsidies?   Government subsidies for forest-management practices must be treated as ordinary income. This causes no problem if the practice is a deductible item. If you spend $100 for TSI work and receive a subsidy of $75, you have deductible expense of $100 and ordinary income of $75, leaving a net expense of $25, which matches your out-of-pocket expense. If you spend $100 for planting, which must be capitalized, you have an out-of-pocket expense of $25 when you pay the planting crew. You will have another out-of-pocket expense of perhaps $35 or $40 when you pay income tax on the $75 subsidy payment. You cannot recover the $100 planting cost until you dispose of the trees planted.

CERTAIN ATTITUDES OF THE INTERNAL REVENUE SERVICE

Occasionally, IRS agents propose to disallow the entire tax loss claimed from tree farm operations, usually in years when there are losses accompanied by no income or very little income. They contend that there is no expectation of profit and that the tree farm is not being operated as a business. I think this attitude of the IRS is due to some court cases it has won. These were primarily cases involving part-time farmers, not tree farmers, who operated the farms as a sideline and in such a manner that the profit potential was remote. I believe you can win such an argument, since the economics of the industry dictate that there are likely to be long periods of time, ranging upward to twenty years or longer, before merchantable timber can be produced. It also appears that the Internal Revenue Code has recognized this matter of timber economics by providing the tree farmer an optional election, even though the option is seldom used, of treating carrying charges and development expenditures either as currently deductible expenses or as additions to capital investment. You may have to prove to the IRS that you are a good tree farmer; forestry finds friends in unexpected places.

The IRS is always anxious that your tax records show the true situation, but some agents might question values you place on bare land, since they usually seem low to everyone except tree farmers. One big trouble is that many people confuse bare land with cutover timberland. Bare land is nothing but dirt; cutover timberland is bare land plus reproduction and perhaps a few scattered trees of merchantable size. As I pointed out earlier, bare land serves the tree farmer only as a base on which to grow trees, and, unless it is exceptionally productive, spending much money for it dooms the enterprise to failure. Much of the money used to purchase cutover timberland goes to buy reproduction, which will eventually become merchantable timber. This is a basic fact of tree farming, and I believe you can win an argument with the IRS on this subject also. You may have to enlist the support of your tax adviser and consulting forester, but IRS agents will usually agree with sound reasoning.

CASUALTY LOSSES AND INVOLUNTARY CONVERSION

Another aspect of taxation important to you is treatment of casualty losses. The most common loss of this nature is from fire; other casualty losses might be caused by windstorm, ice, hail, sleet, unusually heavy rains, floods, freezing, drought, and insect infestation. What problems do you encounter in reporting casualty losses for federal tax purposes?

Loss for tax purposes is determined differently from an insurance loss or the amount of damages for which a negligent party might be sued. Loss for tax purposes is not necessarily your economic loss or decline in fair market value. You must have a depletion basis in timber to incur a deductible casualty loss. I have previously mentioned that this cost basis or depletion basis is composed of purchase price, inherited value, basis acquired by gift, or expenditures capitalized during the period of operations. If loss from casualty is complete, you have no problem in determining loss for tax purposes; it is simply the amount of your timber depletion basis. If there is only partial loss with salvable timber remaining, the measure of loss for tax purposes is the difference between the fair market value of the timber before the casualty and the fair market value after the casualty, with loss being limited to depletion basis. For example, if you have lost through a fire some timber in which you have a depletion basis of $900 and such timber was worth $5000 before the fire and nothing after the fire, you have suffered a $5000 economic loss but a deductible loss for tax purposes of only $900.

To support the amount of loss, an accurate appraisal of damage is needed. It may be made immediately after casualty if there is no doubt as to the death of the trees, or it may be delayed until after the next growing season if there is doubt. Waiting until after the following growing season might require an amended income-tax return, since, in most cases, a casualty loss can be deducted only in the year loss is sustained, not in the year in which the amount of loss is determined. A consulting forester should be employed if the amount is large and likely to be contested by the IRS. You need to know three things to establish a casualty loss:

1. Fair market value before casualty
2. Fair market value after casualty
3. Depletion basis of the timber damaged

Involuntary conversion, a type of loss that differs from those just described but is treated in a similar manner, results from seizure by public authority through the power of eminent domain; it is becoming more frequent because of expanded highway-building programs and such projects as reservoirs, pipelines, and power lines. Seizure by public authority often results in taxable gain rather than loss, however, because of tree farm accounting procedures. Depletion basis (or book value) is usually greatest at the time of acquisition and falls gradually with each timber sale. Fair market value nearly always exceeds book value, and condemnation proceeds are based on the former. If it exceeds the depletion basis, you have a taxable capital gain equal to the excess, just as if the timber or land had been sold in a normal transaction. Condemnations are forced sales in the sense that the seller has no control over timing and price; he may not plan to sell or, sometimes, may be unwilling to sell. If the sale produces a gain, he may be forced to pay taxes by being forced to receive income. The Internal Revenue Code acknowledges the possible unfairness of this and allows you to prevent this gain being recognized and taxed by reinvesting the proceeds, within specified limits, in similar or like property. The specified time limits, generally speaking, begin with the threat of condemnation and end one year after the close of the taxable year in which any part of the gain is realized. If only a portion of condemnation proceeds is reinvested in similar property, a gain is recognized and tax paid to the extent that the amount of gain reinvested falls short of the total gain from the condemnation.

Rules of the IRS regarding reinvestment in similar or like property are relatively lenient, for the term "property of like kind" has been broadly interpreted. It appears to be possible to take condemnation proceeds from a tree farm and reinvest them in other real estate quite different from the timber tract given up, such as ordinary farm, office-building, apartment-house, or unproductive raw land to be held for future development as a subdivision or shopping center, and still qualify for nonrecognition of gain.

This tax provision helps to maintain the price of tree farms, and each condemnation has a nationwide effect. If, as is often true, the property condemned has a very small depletion basis, most of the consideration may be a capital gain. The owner must pay about a third of the money received in taxes, unless he acts within a relatively short period, and he balks at this. As his time gets short, he thinks he can afford to pay more than a property is worth, since he is spending 67 cent dollars. The effect is nationwide because an owner may lose a property by condemnation in Oregon and replace it with one in Pennsylvania. The alert seller should watch the progress of all large projects where condemnation is substantial; he may find a buyer for his property among those who have lost theirs and must spend the money. This is one of the plus factors that cannot be measured.

DEPLETION-BASIS PROBLEMS OF INHERITANCE AND GIFTS

Let us touch briefly on some problems encountered in estate- and gift-tax matters. The first point to keep in mind is that inherited property has a "carry-over" basis in the hands of the new owner. This means that the cost basis in the hands of the decedent carries over and is the cost basis in the hands of the beneficiary. There is one exception. Property acquired by the decedent before December 31, 1976 and which had appreciated in value before that date takes a "fresh start" basis in the hands of the beneficiary. The "fresh start" basis is the value on December 31, 1976. This value is not determined by appraisal, but by applying a statutory assumption that appreciation was spread equally over the time the property was owned. Therefore, if you inherit a tree farm, you should determine from the executor of the estate the amount of the cost basis of the decedent. Determine the value applicable to land and that applicable to timber, and make these values part of your permanent records. Since it may be many years before inherited timber is ready to be sold and, therefore, many years before you need these values to determine your gain on a sale, it is a good idea to obtain this information while all facts are still available.

You may have the opposite problem of passing tree farms to your eventual heirs in such a way that taxes will cause the least possible disturbance. If your tree farm contains much merchantable timber, your heirs can probably sell enough timber to pay the taxes. On the other hand, if your tree farm contains well-stocked stands of trees below merchantable size, estate taxes may be sizable and force the sale of the land itself, since income from other sources is impossible. Under present laws, farms, including tree farms, may be valued as farm property, within certain limitations, rather than as property with a higher use. This is a very technical area, with numerous conditions to be met in order to qualify. Recent tax-law changes have largely eliminated advantages of making life-time gifts to individuals, except for annual gifts of $3000 per donee.

You can give timber property to most charitable organizations and obtain a charitable-contribution deduction on your income-tax return equal to the fair market value of the property at the date of the gift, provided the charitable deduction does not exceed certain limitations. It is often much cheaper, taxwise, to make a contribution of property that has appreciated in value than to make a cash contribution, because appreciation in the value of the property escapes income taxation. Estate- and gift-tax problems are commingled and complicated, and you should not make any moves without careful study with your lawyer, your tax adviser, and possibly a consulting forester.

IDEAS FOR YEAR-END PLANNING

Federal tax laws add another attractive feature to tree farm ownership if you are on a cash basis and report your income by calendar years, a situation that includes almost everyone. You can influence each year's taxable income to a certain extent by tree farm operations. Annual growth is complete by the end of summer, and September is a good time to take stock and make plans. If your income for the year has been substantial, you can schedule necessary deductible expenses for the last quarter of the year. Both TSI work and boundary-line maintenance fit readily into such a program, and you can probably pay the year's ad valorem taxes in late December. You can also speed up charitable contributions. If your year has been unfortunate up to September, you can reverse this process, speeding up income and slowing down expenses.

The flexibility of timber sales is even more helpful. You can prepare the sale in October or November and have bids returned during the last few days in December. When you know the exact amount of the consideration, you can make it fall into either year by hurrying the close or dragging your feet, or you can spread the income over several years by an instalment sale or a deferred-payment arrangement. Selling timber is slower than selling stocks listed on the various stock exchanges, and you need to plan ahead if you want to take advantage of these opportunities.

ACCOUNTING RECORDS

Since normal operation of a tree farm does not include frequent transactions, your accounting records need not be complicated, but some things are essential. Foremost is the cost or depletion basis of your timber. No matter how or when you acquire the property, you must assign some value to trees and some value to land. You can make these assignments merely by a penciled memorandum on your ledger sheet or check stub, but the record should be permanent, since you will often refer to it in later years. If you have failed to do this, any consulting forester can make a formal appraisal of the value of your timber when you acquired it. His records extend for many years and cover many sales, and his report, when properly done, will be acceptable to the IRS.

Also important are records to substantiate operating expenses. Here again, they may be only penciled memoranda of trips you took on certain days to inspect your tree farm, showing mileage and other expenses. Check stubs for other expenses will be sufficient. You will find that memory is not good enough for either your purposes or those of the IRS, and a poor memory costs money in excess taxes. Penciled memoranda may be sufficient as long as they are permanent, but I strongly recommend more formal records. In accounting, as well as in law and forestry, you must make your record as you go or run the risk of trouble. Revenue agents are reasonable but exceedingly thorough men, and you can help your tax adviser help you if you maintain proper records.

The IRS and your tax adviser will insist that your records be detailed enough to show what profits you make from your tree farm, but good records serve other purposes too. What you make on invested capital is of great importance, and your tree farm must perform well in comparison with other possible investments, or you should sell it. Your tax adviser and consulting forester can help you set up a simple means of measuring this performance for your own use. Such a record is likely to be valuable if you ever have to sell the tree farm. Rate of return is important to all investors, and records of actual profitable operations are powerful sales weapons.

11

The Capital Gains Tax

Early in the preceding chapter, I defined what is commonly called "capital gain" as a gain on the sale of a capital asset that has been held longer than 12 months. This gain receives special tax treatment that applies to all capital assets such as stocks, bonds, and so forth. Money invested in these assets is tied up for, and subject to many risks over, what might be years and years. It is not fair to tax as ordinary income in one year at high rates a capital gain accumulated over a long period of time.

Congress acknowledged this in the early 1920s and made only half of the gain taxable at ordinary rates, and it further decreed that the tax shall not exceed a certain percentage of the total gain (at present the percentage varies from 25 to 35). Therefore, if your annual income is low during the year you obtain a capital gain, you can elect to pay ordinary income taxes on half of the gain in order to reduce your tax. Regardless of your income, the tax on capital gains will not exceed 25 to 35% of total gain. You often hear someone say, "I don't want to sell my timber [or tree farm], because taxes will eat up all my profit." This is not true. The rate is most favorable, and the speaker either is complaining about a happy situation or does not know the facts.

Application of this principle to tree farming is uniquely appropriate, since trees are long-term investments. A tree planted today will not be merchantable for 15 years even under favorable conditions. During its long struggle to get to market, it may be destroyed by fire, insects, disease, and weather, and it must bear all management costs described elsewhere. Purchase of a tree farm must be a long-term commitment, since no market for land offers the liquidity of markets for most stocks and bonds. Ordinary income-tax rates rise sharply as annual income increases, until, at the highest level, 70% of each dollar goes for federal taxes. You may sell a tree farm only once in your lifetime, and the gain will cause an abnormal increase in your income in the year of sale. If this gain were subject to ordinary income-tax rates, it might be almost eliminated by taxes.

Prior to 1944, capital gains treatment applied only to outright sales of timber, but, in that year, Congress recognized that this discriminated against good forest-management practices necessary to provide raw material for forest industries. For example, a lumber company providing 100 jobs in its manufacturing plant alone could not afford to buy land and grow timber for its own use, and thereby increase the stability of its operations, since the value of the timber it used in its mill was subject to ordinary income taxes. To qualify for capital gains treatment, it had to sell these trees to someone else. This encouraged the company to "cut out and get out," leaving the community to pick up the pieces. In 1944, therefore, Congress extended the capital gains provision to include timber cut by its owner for his own use.

The outright-sale restriction had also made it impossible for tree farmers to obtain capital gains treatment of timber sold on a paid-for-as-cut basis. Many small trees removed in thinnings must be sold on this basis in order to get any reasonable price, and income from these early sales is almost essential. Timber of very poor quality and defective or inaccessible timber must often be sold on this basis or not at all. Since the old rules made sales of this type subject to ordinary income-tax rates, they discouraged good forest management, and Congress corrected this in 1944 also.

The wisdom of the 1944 Congressional action has been strikingly demonstrated. For eight years prior to 1945, the average annual production of timber products was 5.9 billion cubic feet, while the volume of our timber fell 9%. During the eight years after 1944, the average annual production of timber products was 7.0 billion cubic feet, while the volume of our timber rose 2%. Annual tree planting on private lands rose from 200,000 acres in 1942 to over 1,200,000 acres in 1959. From fears of a timber famine in 1942, we have moved to an economy of relative abundance, where timber grown exceeds timber cut by a comfortable percentage; this is an impressive accomplishment in such a short time. Nevertheless, the balance is delicate, and we need a firmer base from which to meet the staggering demands of the future.

What Congress can do, it can undo, and there is a constant effort on the part of some agencies to plug what is incorrectly called "this loophole." "Loophole" is an incorrect term, because it was not an unintended benefit that slipped by unnoticed. It was carefully and deliberately worked out by Congress, and the present rules fit the economics of tree farming extremely well. The elimination of capital gains tax treatment of tree farm sales would endanger millions of small tree farmers for whom this book is written and would almost certainly force most of them out of business. The fight to maintain this provision must be led by associations that are organized for the task; you should do everything in your power to help, and your association can suggest appropriate actions.

Now let us move from the general to the specific and discuss the effect of capital gains treatment on several common types of tree farm operation. First, let us continue the discussion of the investor who made the financial forecast described in Chapter 9 and bought the tree farm for $60,000. For convenience, we shall call him John Bradley.

JOHN BRADLEY—SUSTAINED-YIELD OPERATOR

Bradley first calculates his depletion basis from information contained in the tree farm inventory. After consulting his tax adviser, he assigns $20 per acre ($8000) to land and minerals, $20 per acre ($8000) to reproduction, $30 per acre ($12,000) to pulpwood, and $80 per acre ($32,000) to sawtimber. This establishes the depletion basis for each product.

In his initial appraisal of the tract, he estimated annual income and expenses as follows:

Gross timber income	$7000	
Less selling expenses	700	
Net timber income		$6300
Management costs:		
Taxes	240	
Boundary line maintenance	60	
Total management costs		300
Net income		$6000

His tax adviser tells him that selling expenses must be used to reduce amount of sales, but that taxes and boundary line maintenance are deductible expenses. His trees grow each year; sawtimber trees put on more volume, pulpwood trees grow into sawtimber, and reproduction trees grow into pulpwood. To be completely accurate, each year he should move some of his depletion basis in reproduction up to pulpwood and some of the basis in pulpwood up to sawtimber. If his sales are all sawtimber, he should deplete his sawtimber account by whatever percentage the amount sold is of total volume present. For our present discussion, however, we shall assume that he will sell all sawtimber in the first 10 years and that his depletion charge each year will be 10% of the basis, or $3200. Therefore, since his first sale will occur one year after purchase of the tract, he will have average annual sales of $6300 less average annual depletion of $3200, leaving average annual capital gains of $3100.

Since he borrows all the purchase money (by pledging the tree farm as security and from other sources available to him), the biggest single annual expense is interest on the unpaid balance of his loan. This is deductible without doubt in most cases.

Bradley has a moderate income from other sources and elects to divide his annual capital gain by two and add this to his other income for tax purposes. Therefore, his taxable income from tree farm operations is only $1550. Each year, he pays interest and part of the principal of his loan, and these figures can be seen in Table 9–1 (page 84). Table 11–1 shows how he stands on the last day of each year for the first 10 years.

Now let us look into the future. Bradley thinks he can sell the tree farm at the end of 10 years for just what he paid for it, $60,000. His depletion basis is then only $28,000 (the original total less sawtimber depletion), so he has a capital gain of $32,000. Federal tax on this gain varies with the size of his income, but the maximum is $8900. The tree farm is located in a state that does not recognize capital gains treatment of any income, so he will owe state income taxes of 4% on the entire $32,000, or $1280. His total tax bill is $10,180. Subtraction of this from the sale price of $60,000 leaves $49,820. He still owes $42,615 on his loan, and payment of this leaves a remainder of $7205. This will be all his, but it seems a small amount for 10 years of work and risk. Nobody would seek such a business venture for such small returns.

TABLE 11–1

Year	Loan Balance	Sawtimber Depletion Basis	Income (Half of Capital Gain)	Deductible Expenses Management Costs	Interest	Tax Loss (−) or Gain (+)
1	$58,800	$28,800	$1,550	$300	$4,800	$−3,550
2	57,504	25,600	1,550	300	4,704	−3,454
3	56,104	22,400	1,550	300	4,600	−3,350
4	54,592	19,200	1,550	300	4,488	−3,238
5	52,959	16,000	1,550	300	4,367	−3,117
6	51,196	12,800	1,550	300	4,237	−2,987
7	49,292	9,600	1,550	300	4,096	−2,846
8	47,235	6,400	1,550	300	3,943	−2,693
9	45,014	3,200	1,550	300	3,779	−2,529
10	42,615	0	1,550	300	3,601	−2,351
						−30,115

The incentive lies in the tax losses Bradley sustains each year without spending any additional cash. Refer to the last column in Table 11–1. His tax loss varies each year with the amount of interest he pays, but the average is $3012. Although Bradley does not have a large income, he pays federal and state taxes of 25% on some of it. His ownership of the tree farm permits an annual tax saving of 25% of the $3012, or $753. This plus a tenth of the anticipated profit of $7205 means an average net cash income, after taxes, of $1474 annually. This will not make him wealthy, but do not forget that he borrowed all the money at 8% or at least made all invested capital, regardless of its source, earn an 8% return before determining his profit.

In addition, he has a chance at big profits. Suppose growth of the nation, expansion of timber markets, and a steady slow rise in the price of land make it possible to sell the tree farm for $120,000 at the end of 10 years. This is a likely possibility. Repetition of the tax calculation for this price shows that his net cash profit after taxes and payment of the loan balance would be $45,325. During the period, he will have a chance at extra profits from minerals, which would be especially helpful, since a percentage of this income will be depletion, although the remainder will be taxed at ordinary income rates. He will also have a chance at the items of miscellaneous income covered in Chapter 5. Although most of these represent ordinary income, any income is valuable, because even the highest tax rates leave something.

This cannot go on forever, since the depletion basis cannot exceed actual cost and it is already almost eliminated. Let us see what will happen during the next 10 years. The only depletion basis Bradley has left in merchantable timber is $12,000 in pulpwood. Let us assume that he will liquidate all of this during the second 10 years. Therefore, during this period, his annual depletion is $1200, leaving an annual capital gain of $5100, half of which is taxable. Management costs remain the same, and interest depends upon the loan balance. Table 11–2 shows his position on the last day of each year of the second 10-year period.

You can see that he will eventually make a profit, but only a small one. Twenty years from purchase he will have exhausted his depletion basis and will have remaining only that in reproduction, which by then should be moved up into timber, and land and minerals, which he cannot recover until he sells the tree farm. He will have almost retired his loan and will have reached the point where a few hundred dollars in annual sales will carry the property. Most investors know that 20-year predictions are almost worthless. I carried out this calculation just to show what would happen.

Although tree farming is normally a long-term venture, it is also attractive for the short term. To illustrate this, we shall now consider the desires and plans of Thomas Stewart.

TABLE 11-2

Year	Loan Balance	Pulpwood Depletion Basis	Income (Half of Capital Gain)	Deductible Expenses Management Costs	Interest	Tax Loss (−) or Gain (+)
11	$40,024	$10,800	$2,550	$300	$3,409	$−1,159
12	37,226	9,600	2,550	300	3,202	− 952
13	34,202	8,400	2,550	300	2,978	− 728
14	30,938	7,200	2,550	300	2,736	− 486
15	27,413	6,000	2,550	300	2,475	− 225
16	23,606	4,800	2,550	300	2,193	+ 57
17	19,494	3,600	2,550	300	1,888	+ 362
18	15,054	2,400	2,550	300	1,560	+ 690
19	10,258	1,200	2,550	300	1,204	+1,046
20	5,079	0	2,550	300	821	+1,429
						+ 34

THOMAS STEWART—SHORT-TERM INVESTOR

Stewart is 56 and has an income so large that many of his dollars are subject to federal taxes of 50% and state taxes of 4%, allowing him to keep only 46¢ out of each dollar. He is in his peak earning period and is anxious to accumulate funds to make his later years more comfortable. Therefore, he is concerned with net income after taxes. Certain aspects of tree farming meet his needs very well.

Stewart is offered an 800-acre tree farm from which all sawtimber and most pulpwood was removed seven years ago. The price is $100 per acre and includes half the minerals. Although Stewart is not a tree farmer, he decides to investigate this offer with the help of a consulting forester. The first step is a tree farm inventory, even though there is obviously little merchantable timber. The inventory reveals merchantable timber worth $12,000, an insignificant amount, but also shows that the entire tract is covered with young trees of desirable species. These trees are the children of those removed seven years ago, but they are over-topped by undesirable trees that are not merchantable. The forester states that removal of these undesirables will produce a dramatic response in the trees underneath them and estimates the cost of this TSI work at $16,000. This expense is deductible, a magic word for Stewart, and he continues his investigation.

Although the report does not include a detailed soil analysis, the forester's investigation shows that the trees cut earlier were tall and vigorous and that trees now on adjoining lands are equally so. This indicates that the soil is quite productive or at least that there are no serious problems. There are no signs of recent fire on or near the tract. The boundary lines were surveyed and painted two years ago, and there is no evidence of trespass or adverse possession. The property fronts on a good highway for half a mile and is within 35 miles of a proposed paper mill. It is nine miles by highway from a pulpwood yard operated by an existing paper mill about 90 miles away, and purchases at this yard have been steady for four years. The market for sawtimber of all kinds is good. Ad valorem taxes are $480 annually. None of the land appears likely to be in demand for conversion to higher use in the near future. Nevertheless, several quail-hunting ranges managed by professional dog trainers are nearby, and interest in this sport is high. These facts convince Stewart and his forester that TSI work now will enable Stewart to sell the property for $200, possibly $250, per acre six years from now. Stewart consults his tax adviser, and they work out a comparison of his situations with the tree farm and without.

Calculation of depletion basis is the first step. Stewart plans to cut no timber during his ownership; therefore, assignment of separate values to timber and to land is not essential initially and can always be done later by using the tree farm inventory and the records of the forester. The cost of the inventory is $1200 and must be capitalized in this case. Therefore, his depletion basis is the cost of the tree farm ($80,000) plus the cost of the inventory ($1200), a total of $81,200. Stewart's financial position enables him to borrow the entire purchase price, provided he spends $16,000 on TSI work within three months after acquiring the tree farm, and he agrees to do so. He must pay 8% interest on the loan, annual ad valorem taxes of $480, and $250 annually to the consulting forester for advice, general supervision, and administration. The TSI work, interest, taxes, and administration are all deductible expenses.

Stewart can forgo the purchase, pay taxes on the income that would be used for tree farm development expenses and carrying charges, and use the remaining funds for other investments. For comparison, he must determine how much of the planned tree farm expenses will go for taxes, how much will be left over for investment, and what the other investments will produce in the same period of time. He thinks that the most attractive alternative is a tax-free municipal bond paying 6%, and he can get compound interest on the bond by reinvesting the interest as soon as it is received. Table 11–3 shows the amount of money he will have six years from now if he chooses the bond.

If his original projection is correct, he will sell the tree farm at the end of six years for $160,000. The sale price minus the depletion basis of $81,200 will leave a capital gain of $78,800. This is how much cash he will have after paying off his loan of $80,000 and deducting cost of the inventory. The

TABLE 11-3

Year	Deductible Expenses			Amount He Can Keep	Interest Factors to End of Period	Cash after Taxes at End of Period
1	TSI work		$16,000			
	Carrying charges					
	Interest	$6,400				
	Taxes	480				
	Management	250				
	Total		7,130			
	Total deductible expenses		23,130 × 0.46 =	$10,640 ×	1.338 =	$14,236
2	Total carrying charges		7,130 × 0.46 =	3,280 ×	1.262 =	4,139
3	Total carrying charges		7,130 × 0.46 =	3,280 ×	1.191 =	3,906
4	Total carrying charges		7,130 × 0.46 =	3,280 ×	1.124 =	3,687
5	Total carrying charges		7,130 × 0.46 =	3,280 ×	1.060 =	3,477
6	Total carrying charges		7,130 × 0.46 =	3,280 ×	1.000 =	3,280
						32,725

federal tax is $26,797, and the state tax is $3152. Therefore, the remainder after taxes will be $48,851, $16,126 more than he would have if he bought the bond.

Remember that Stewart visualized a possible sale at $250 per acre. If this price is obtained, the remainder after taxes, using the same calculation, will be $71,301, more than twice the amount produced by the bond. Stewart buys the tree farm. He feels that a small profit is certain and that he has a chance at a large profit. In addition, mineral ownership and the chance to negotiate recreation leases add extra appeal to ownership.

This illustrates that tree farming can be attractive to investors whose interest is only temporary and who do not want to plan ahead more than a few years. Stewart, however, is valuable to the economy, for he takes the first step in providing raw material for forest-products industries and also lays the groundwork for profitable operations by the next owner. Let us look six years ahead and consider the desires and plans of William Kimble, who buys Stewart's tree farm.

WILLIAM KIMBLE—YOUNG FAMILY MAN ON THE RISE

Kimble is 37 years old, married, and has three children, aged 12, 7, and 5. He has worked 15 years for the same company and receives a salary sufficient for comfortable living and some capital accumulation, and he is anxious to put aside something for the college education of his children and for his eventual retirement. He is offered Stewart's tree farm at $250 per acre and decides to investigate.

His consulting forester, even at first glance, realizes that the tract is quite valuable and elects to make an intensive inventory and examine closely as much of it as possible. He finds a solid stand of desirable trees about 13 or more years old, just below merchantable size, with a sprinkling of larger trees. There is no evidence of fire, trespass, or adverse possession, but the paint on the boundary lines is dim. The trees form a solid canopy, needing thinning as soon as possible, but cannot be thinned for the first time until five years from now. Although growth has begun to slow down, it will remain adequate until the trees are large enough for pulpwood. His growth study shows that a thinning five years from now will produce $56,000 after all selling expenses. It also shows that it will be possible to clearcut the stand 13 years from now for a net realization of $449,000. Ad valorem taxes are $480 per year, and management costs are estimated at $400 per year, since boundary line maintenance is necessary. The timber stands are so thick that there are few possibilities other than tree farming. The forester's fee is $1800.

Kimble then sits down with his tax adviser to calculate the depletion basis. The price of the tract is $200,000, which, added to the forester's fee, makes a total cost of $201,800. They allocate $20,000 to land and minerals and $181,800 to timber. They determine that the thinning five years from now will be all depletion and that the harvest cut will be part depletion and part capital gain.

Next they look at carrying charges. Kimble can borrow the purchase price of $200,000 at 8% (by the same method Bradley used), making annual interest cost $16,000. Ad valorem taxes of $480 and management fees of $400 increase annual carrying costs to $16,880, and these are all deductible expenses. More than this amount of Kimble's annual income is subject to combined taxes of at least 40%, so he is required to invest only 60% of this, or $10,128, from funds he could use for other purposes. He and his tax adviser think the best alternative investment is a high-grade common stock that, by a combination of dividends and appreciation, will enable him to obtain an after-tax return of 7% compound interest. Kimble calculates that this alternative program will produce a total of $58,247 five years from today, just about the time his first child starts to college. Since the first thinning will produce an after-tax sum of $56,000, the investments are nearly equal at this point, and common stock is preferable because of its liquidity. They continue the investigation.

The predicted harvest cut produces a large sum of money, but it will not be all his. First, he has taxes to pay. His timber depletion basis is now $125,800, down from $181,800 because of the thinning mentioned above, so he has a capital gain of $323,200. Application to this of the appropriate federal and state tax rates gives a tax bill of $137,179. Second, he must repay his loan of $200,000. Subtraction of these two items from the $449,000 leaves $111,821 in total net cash. He compares this with results produced by the alternative program. The same calculation as that described above, only for eight years this time, shows that the common stock will produce an after-tax sum of $103,925. The tree farm is the better deal by $7896.

In addition, he still has the land and minerals with a depletion basis of $20,000. If he can sell them for this price, he pays no taxes, and the tree farm looks better by an additional $20,000. Although conditions thirteen years from now are hard to predict, he thinks that a sale at this price is a certainty and that he has a good chance to get three or four times as much.

Although common stock has the advantage of liquidity, the tree farm has several plus factors that cannot be measured. First, the mineral interest may produce lease bonuses and rentals and may even rise spectacularly in value. Second, technological progress may cause a rise in price of tree farm products. Third, inflation may do the same thing. Fourth, the expanding population may make all land more valuable.

These considerations make Kimble interested in buying the tree farm, but he still thinks that the estimated profit of $7896 plus the land and minerals is not enough in proportion to the risks involved in borrowing $200,000 and operating the tract for 13 years. Since the interest cost is such a huge factor, he recognizes that the tree farm would be much more attractive if he could buy it for less. Therefore, he makes a counter offer of $160,000. Stewart in turn counters with $180,000, and Kimble buys the tree farm.

DISCUSSION OF BRADLEY, STEWART, AND KIMBLE

Possible variations on this theme are almost endless and, of course, the calculations above are much simplified to make my point, which is that tree farming is a good investment for people with widely differing objectives. These three men are alike only in that they have money to invest, and yet all three made money. I point out also that, although the greatest profits come at the end of a long term, you can make money over a short term in a transaction of an investment nature (as opposed to speculation). Stewart does well, even though he is in the business only a short time. All three may make big profits if some of the unpredictable factors work in their favor.

The attractiveness of each investment depends heavily on capital gains tax treatment of income, and I urge caution here. Although all timber sales can be made eligible for this treatment, they do not automatically come under this provision. Stewart will have no trouble, since he buys land, holds it for six years, and then sells everything without cutting any timber.

Bradley and Kimble, because of their timber sales, may be classified as men who grow or hold timber for sale to customers in the ordinary course of business, and the profit on their sales may be ordinary income. In order to avoid this, they may sell timber under a contract by which they retain an economic interest in the timber, such as might exist in a pay-as-you-go sale. Either suffers a severe blow if profits are taxed as ordinary income. For this reason alone, timber-sale contracts are complicated and important matters on which expert advice is mandatory.

These examples show three prudent investors using the proper approach. You can follow their examples with profit. At the moment of decision, probably the most important moment in tree farming, they used the best professional advice available and were exceedingly thorough at every step.

SELECTION AND USE OF YOUR TAX ADVISER

The tax laws applicable to tree farms were complicated enough for most laymen when I first wrote this book in 1964. Major complications were added in 1976. For example, a change in so-called "tax-preference" items has caused the imposition of a minimum tax, which sometimes will reduce the benefits of favorable capital gains taxes. Long-term gain on land and timber sales is a tax-preference item. In a separate computation, one-half of net long-term capital gain in excess of short-term capital loss is reduced by the greater of (1) $10,000 or (2) one-half of the income tax due for the year. The excess is taxed at 15% in addition to income tax otherwise determined.

More such complications will surely follow. On the other hand, income taxes are a major problem only at times of timber sales, perhaps at intervals of ten to twenty years. Therefore, it seems to me that, just before and after timber sales, professional tax help is a must.

A well-known tax adviser is the certified public accountant, and other accountants and some lawyers may be equally competent in this field. Any man active in business can help you locate one. You need one familiar with tree farm operations, and these are fairly numerous in timbered sections. Any adviser can read tax laws, but he must base his recommendations on records you furnish him and things you tell him. If he is familiar with tree farming, he will remember to ask you for items you forget, such as deductible expenses, and he will be much better at suggesting possible tax advantages in all tree farm operations.

Discussions in this chapter and the foregoing one are based mainly on the federal tax treatment of tree farm income and expenses, and, although states generally use the same rules, there may be important differences. State laws often require that you pay income taxes to the state in which the tree farm is located, even though you may live elsewhere; payment of such a tax usually allows a credit against the tax you owe the state of your residence, so double taxation rarely occurs. Therefore, it is wise to consult a tax adviser in the state where your tree farm is located, and this is especially important in case of death of nonresident tree farmers, since estates will probably have to be administered in both states.

These two chapters give you an outline of the tax treatment of tree farm incomes, and they point out some possibilities open to you. Taxes are constantly changed, however, and you must consult your tax adviser continually. Tree farming is a business where pennies count, and no amount is unimportant.

12

The Lawyer

Law reaches into the daily life of all of us, and a lawyer is more important in tree farming than is generally realized. Many people seek a lawyer only when they get into trouble, but he is most valuable in keeping you out of trouble. Tree farming puts a high premium on the latter because of its long-term nature and because mistakes may come to light many years after they happen and much too late to correct. A tree farmer without a lawyer is almost certainly headed for trouble.

The lawyer's services to you fall into two main divisions, those in connection with purchase and sale of tree farms and those in connection with their operation. Let us discuss them in this order. The first division is a two-way street that you will travel in different ways at different times.

IMPORTANCE IN PURCHASE AND SALE OF TREE FARMS

Purchase and Sale Contracts. Your lawyer prevents trouble when you buy or sell by making sure that you reach definite agreements on every aspect of the trade. A written contract may be necessary only when large amounts are involved or when there is likely to be a long time between the first agreement to buy and sell and actual closing of the trade. In any case, however, it is essential to decide on all details, and, unless the sale is actually closed at the time of the first agreement, some written memorandum is almost mandatory. Enforcement of an agreement to buy or sell by either party is virtually impossible without something in writing. Timing is most important, because people are human. Both parties to a trade are in their most agreeable humor at the time of the original agreement, and it is easy to make these decisions then. As time passes, each assumes that unsettled matters will be settled in a manner favorable to him, and both parties may change their minds. This human characteristic sets the stage for disagreements, irritating and unnecessary at all times and sometimes costly. A lawyer is an expert in avoiding disagreements, and he eliminates the causes of many by asking the following questions:

What is the consideration, and how is it to be paid? Cash payment is relatively simple, but many sales are made on an instalment basis. An instalment sale requires, in addition to a deed, preparation of a note and a mortgage, deed of trust, or other security instrument.

What is to be conveyed? Do any reservations arise from this trade or from some earlier time? Mineral reservations are common, and timber, grazing rights, ROW's, and many other things may be reserved or excepted; reservations may be either permanent or temporary. What form of conveyance is to be used? Further discussion of quitclaim and warranty deeds appears later.

What kind of title information is required, and who is to furnish it? The next section deals with title work in detail, but you must agree first on what will be done and who will pay for it. Ordinarily, the seller furnishes evidence that he owns what he is selling, but this feature may be the subject of negotiation.

Who is to pay current ad valorem taxes? The seller often pays the portion that has accrued up to the date of the sale, but they may be assumed by the buyer. If they are prorated using the amounts for the previous year, the simplest solution is for the seller to give the buyer a check for the pro rata share of taxes at the time of closing and recite in the deed that the buyer agrees to pay all ad valorem taxes for the year. There may be a problem here if the sale includes only part of the seller's land; the tax collector may insist upon payment of all taxes or none.

Who pays for any state tax stamps that must be affixed to the deed? These are usually the responsibility of the seller, but they may become the obligation of the buyer if the seller fails to affix them. Many states require tax stamps of one form or another, particularly if some of the minerals are separated by the deed.

Is a commission to be paid? To whom and by whom? Commissions are usually paid by the seller, but they may be negotiated. An attempt by the responsible party to avoid paying a commission that has been earned may cause unpleasantness for both parties.

How and where is the trade to be closed? The two parties can meet anywhere and exchange the deed for money, but more indirect methods are usually selected. Often it is sufficient for the seller to deposit the deed with a bank, accompanying it with a letter stating to whom and under what conditions it is to be delivered. A real estate broker or lawyer may serve as escrow agent under a simple agreement when the transaction includes earnest money. As the transaction increases in size and complexity, more formal arrangements are necessary.

What sort of recording is necessary and who pays for it? The buyer usually records the deed at his own expense, since this is public notice of his purchase, but the seller may require recording to show his title to certain reservations and his fulfilment of requirements about tax stamps. Mortgages are usually recorded by the beneficiary. A more satisfactory arrangement may be to include recording of all instruments among the services of the closing attorney or agent. If a mortgage or deed of trust arises from the sale, this method is especially desirable, since the beneficiary wants to be sure that the mortgage is a first lien.

Who prepares the necessary papers and who pays for this? Although others may be willing, this is certainly a job for a lawyer, and, the more experienced he is, the better. In few other activities does the do-it-yourself method bring such risks of disaster. Very few lawyers get rich from handling these details in an expert manner; some have gotten rich in extensive litigation necessary to correct errors of ignorance at this point.

The questions raised above indicate the important parts of purchase and sale agreements, and the many details may seem overpowering. In the normal transaction, however, they are not, and they can often be settled in less time than it takes to describe them. They must all be answered, for there are at least two solutions to each. Although the law will settle those not settled by the parties, this solution may not be satisfactory to either, and everyone is usually happier over a complete agreement. The conference at which the first agreement to buy and sell is reached is a horse-trading session of the first degree, and the party with the most expert lawyer at hand often comes away with sizable benefits. Thorough preparation for such a conference pays handsome dividends. After agreement has been reached, the next step is to see what the seller owns and the buyer receives.

Title Work.   Almost all of your land was owned at one time by the United States. The federal government acquired it through the American Revolution, purchase, conquest, or treaty. It then passed it along either directly to individuals or indirectly by conveying it to states, which later conveyed it to individuals. Original transfer of title from the United States and all subsequent transfers from one party to another are accomplished by legal documents that are, or should be, recorded in a public office set up for the purpose by the county or other legal subdivision in which the land is located. This recording is public notice that the owner has acquired the property. The transfers from one owner to another make up the chain of title, and, under ideal conditions, each transfer from the United States down to the present owner is on record. Such a chain of title shows who is the record owner of, or has record title to, the property. Record title is a strong claim to ownership, and this is what a lawyer investigates in his title examination. The other type of claim to ownership is based upon possession and may be an even stronger claim.

After the lawyer finishes his title work, he reports what he has found and gives his opinion of the title and its merchantability together with the basis therefor. His opinion may be based on an abstract, a summary of the important provisions of every paper on record showing transactions that affect title to the tract. An abstract is often quite bulky and expensive, and your lawyer can tell you whether it is necessary or desirable. Methods of making title examinations and types of public records vary from state to state.

The lawyer's opinion is what really matters, and you must read it carefully. He may say that the owner's title is virtually worthless. In the usual form, the lawyer states that he finds the owner has a good title subject to certain exceptions, and then he lists these exceptions. One exception almost always found is the ad valorem tax for the current year, which is not yet due and payable. Similarly excluded are facts that a survey of the property would reveal. He looks only at public records and is unaware of adverse possession or boundary-line disputes that exist on the ground. The tree farm inventory should reveal these. He will also show the existence of reservations or easements but usually will not give all their provisions. You should get a copy of these instruments and read them. The clerk of the office where title records are kept can provide one for a nominal fee. Study of these instruments is essential; fine print in one of them may eliminate your chance for profit. There may be other exceptions that pertain to previous financial transactions not properly closed out. Some lawyers specialize in title examination, and their ability to spot a title flaw is uncanny.

Once the flaws have been determined, discuss them with your lawyer to see how serious they are and what can be done to correct them. Many technical flaws may be so inconsequential that they are not worth the expense of correcting. Some serious flaws may require expensive corrective action to eliminate them completely but can be reduced greatly in importance by simple expedients. Others may render the title worthless. The discussion should also cover those facts revealed by the tree farm inventory or other work done on the ground. The time to explore these matters is before you buy the property, not after.

In order to continue our discussion, we must digress momentarily to consider the nature of the statute of limitations and a few differences between quitclaim deeds and warranty deeds.

As an illustration of the statute of limitations, suppose that you punch John

Jones in the nose. You will make him angry and may send him to the hospital for expensive repairs on his nose and, perhaps, his dignity. He may stay angry forever and can do so as long as he wants. He may also want to sue you and to recover the cost of repairing the damage you did. He probably has the right to sue you, but he must do so within a limited period of time. The length of this period is set by a statute of limitations and varies with the type of action and the state laws governing it. In everyday language, the statute of limitations begins to run the day you punch him. He must sue you within the specified period, or he loses his right of action against you. This kind of law makes it impossible for Jones to keep you in permanent jeopardy because of this one event.

Statutes of limitations are important to the tree farmer because they apply to such things as timber theft, adverse possession, and damage of all kinds. Most people hate arguments and disputes and, for this reason, often delay action when they catch a timber thief or suffer a loss from fire caused by a neighbor. Postponement of a decision is a decision in itself. You must either do something about these things promptly or lose your right. In many cases of poor management, loss is not even discovered until the time limit set by statute has already expired. Existence of these laws is one more strong argument for good tree farm management, which requires the services of a capable lawyer to interpret applicable statutes.

A quitclaim deed conveys whatever interest the signer, or grantor, owns. The grantor may own everything or nothing, and he makes no guaranties to the buyer, or grantee. In a warranty deed, the grantor conveys whatever interest he owns, but he also guarantees, among other things, that the title is good. This does not mean that the title is good; it merely means that grantor guarantees it is. Obviously, the guaranty is no better than the man who makes it, and ordinarily his liability is limited to the amount of the purchase price. Suppose that, unknown to the grantor, a flaw that later causes title failure existed at the time of conveyance so that he could not have conveyed a good title. The statute of limitations, in some instances, begins to run on the date of conveyance, and the warranty may be barred before title failure is known. These facts may render a title worthless; a quitclaim deed to a good title may be better than a warranty deed to a bad one. This brief discussion barely touches a complicated subject but does point up the advantages of title insurance.

Title insurance is issued by many private companies and provides compensation up to the face amount of the policy for loss sustained by title failure. You pay the premium once only and are covered as long as you own the property. You must consult your lawyer for all the details of title insurance; I recommend it for many reasons. First, the statute of limitations does not run on its coverage. Second, it protects you against loss from factors that title examination cannot reveal, such as insane persons or forgeries in the chain of title. Third, the insurer agrees to defend your title against all lawsuits, even those of a nuisance nature only. Fourth, some policies protect you against loss from lack of a right of access to and from the land, and this can be very important to you. The cost of title insurance is usually less than one-third of 1% of the insured value, and the premium must be capitalized, not deducted. The decision to obtain title insurance is the buyer's, but responsibility for its cost is a matter for negotiation. Title insurance covers conditions when the policy is issued; consequences of your later actions are your responsibility.

Three other points are worth mentioning in connection with your record title. First, it is always possible that public records will be destroyed by fire or some other disaster. In such a case, your record title can vanish in a moment. Therefore, you must keep in a safe place all documents pertaining to your title. They will support your claim if something happens to the public records, and they may have to be recorded again at some later date. Title insurance is helpful in such an event.

Second, possession will eventually cure almost any record-title defect but is not effective against any government. The document transferring title from the United States directly to an individual is commonly called a "patent," and it, or a certified copy of it, should be obtained and recorded. It is usually an impressive document complete with ribbons and seal and can be obtained at nominal cost from the Bureau of Land Management of the Department of the Interior, in Washington, D.C. If title to your land passed through a state, you should obtain and record a document of similar nature from the state agency in charge of land records. Obtaining these patents often requires three or four weeks and may be essential if you have to sell your tree farm. It is also possible that some government still owns your land. Therefore, it is a good idea to take care of this detail now.

Third, you may have acquired your tree farm by inheritance, gift, or other means that made title examination unnecessary; if so, I recommend that you examine your title now. Although some flaws are removed by the passage of time, many are not, and they are most easily corrected when there is no urgency such as there might be when you agree to sell your tree farm or when the statute of limitations is about to run against you. Mineral discoveries likewise make good titles invaluable. Future events are hard to predict, and you can be ready for almost anything if your title is clear.

IMPORTANCE IN OPERATION OF TREE FARM

Timber-Sale Contracts. Sale of timber is perhaps the most frequent tree farm activity. Sale contracts must convey the timber to be sold in a legal and workable manner and contain specific penalties for violations such as cutting of trees not included in the sale. They must also contain language that clearly indicates that the people harvesting the timber are not your employees, unless you fully understand the consequences and are willing to accept them.

Some agencies, governmental and otherwise, have published sample conveyances that might be used in selling timber, and you may be tempted to use them by filling in the blanks and changing words here and there. Do not, except with help from your lawyer. Terms and conditions vary widely from sale to sale, and each is important. Sales may be complete liquidations or partial cuts governed by marking, species, or diameter limits. Lump-sum sales usually provide for cash payment in advance; pay-as-you-go sales require that the conveyance specify exactly the method, rate, and time of payment. All sales have a time limit, but there is often a provision allowing the buyer to purchase additional time for more money (or a higher rate per piece in pay-as-you-go sales). Clauses to cover damage to fences, buildings, roads, livestock, and so on, are common, and responsibilities of the parties in the event of forest fire may vary widely. Either the seller or the buyer may be an individual, partnership, or corporation, and there are small but important differences in state laws applicable to timber sales. These factors make use of sample conveyances dangerous and use of a lawyer highly desirable or mandatory.

Occasionally, you will find a buyer who insists upon using a conveyance drawn by his company or lawyer. It may be quite satisfactory, and you should make every effort to please your customers. Nevertheless, you must never sign such a conveyance until it has been reviewed by your lawyer. This review usually takes only a few minutes and may reveal significant omissions. The buyer may not intend to mislead you; he may just be unfamiliar with your particular situation and is not concerned with your interest or problems.

Right-of-Way Agreements. This is such a big field and such an important source of tree farm income that I have covered ROW's in great detail in Chapter 5. Study of that chapter shows that agreements of this kind are infinitely varied and may have serious consequences for you. You must never sign even the most innocent looking instrument until it has been reviewed by your lawyer. The time to spot pitfalls is before you sign the agreement, not after; your signature may wipe out a valuable trading position. It is hard for me to overemphasize the importance of ROW agreements. You may receive only one request in your lifetime, and you will not have a later opportunity to correct any mistakes you make. The information presented in the first part of Chapter 5 was obtained by long, and sometimes costly, experience.

Public Liability. Since a tree farm is not as busy as a city street, you are not exposed to many risks from a public-liability standpoint, and your exposure usually depends on the status of the people who use your land. If you invite a friend to hunt or fish on your tree farm, he becomes an invitee. If he asks and receives your permission or if he enters without permission, he is either a licensee or trespasser, respectively. You may have certain responsibilities to invitees and entirely different ones to the others. Therefore, it pays to be careful about what you say, and signs that read, "Hunters Welcome," for example, may change your exposure greatly. The predicted growth in outdoor recreation will cause much greater use of tree farms, and you should determine what risks are involved before use starts.

A good solution to this and other related problems is to paint your boundary lines and put up signs that read, "Posted, No Trespassing" at intervals along the lines. Doing this helps to establish the status of the people using your land. It also is a strong indication of possession and may be the only overt kind of possession you can exert until your trees are large enough to cut. Your lawyer can tell you about the proper wording and spacing of these signs.

Many cultural operations such as TSI work or controlled burning are performed by your employees or independent contractors. You have certain responsibilities for their actions, depending on their relationship to you. Use of your employees may make it necessary to carry workmen's compensation insurance, and your public-liability exposure may be much greater, in addition to expanded accounting burdens. Although many of these problems are removed by use of independent contractors, there is usually a remainder of exposure to liability; certain risks are so hazardous by their nature that they cannot be delegated. You can usually obtain adequate protection by requiring the contractor to carry insurance that covers you and him. Your lawyer can list the risks that should be covered, and the contractor's insurance agent will certify that the contractor carries the necessary insurance.

Other Operating Considerations. Your lawyer can help you get a good title when you buy a tree farm, and you will also need his help in maintaining it. Many apparently insignificant actions can affect your title. As purchaser, you can often correct adverse possession before acquisition, but it may start during your ownership. Boundary-line disputes and adverse possession are intermingled problems, and neither is solved by the passage of time. Certain easements can be acquired by adverse possession. On a well-operated tree farm, the owner carries out many different activities from time to time, and these may have legal importance as evidence of possession. Some may be necessary for legal reasons alone. There is always the risk that your tree farm will be damaged by fire or other acts caused by someone else, and the proper action is almost entirely governed by legal considerations. I strongly recommend that, in case of fire, you move rapidly to investigate the cause and origin of the damage. Many lawsuits over fire damages have collapsed in the middle of the trial because the landowner could not prove how and where the fire started. These are complicated problems that require individual solutions and expert legal advice.

SELECTION OF THE LAWYER

The foregoing discussion indicates some of the complexities of your problems. Even apparently insignificant actions may become important, especially because of the long-term nature of your business. Therefore, you need the best lawyer you can find, and you should consult him about the legal aspects of all your actions. Many people are familiar with some legal principles, and you can usually get free legal advice from everyone you know. Do not fall into this trap. Law is a vast field, and every problem requires an individual solution. In this book, I cannot solve, and do not attempt to solve, the legal problems that face you; I can merely call your attention to the importance of your lawyer.

Finding him should be no problem. This time you need a specialist in problems of tree farmers and one with the widest possible experience. It is helpful if he is a tree farmer himself; any consulting forester can give you names of several lawyer–tree farmers. Nearly every law or legal principle mentioned in this book varies from state to state, so you certainly need a lawyer familiar with the laws of the state in which your tree farm is located. If you live in another state, your local lawyer can help you locate one.

Lawyers have much the same fee arrangement as consulting foresters, and it is wise to discuss this early. In many cases, your lawyer can give you the exact amount of the fee, and, in others, his estimate is usually close to the final figure. He usually charges in proportion to what he does and how long it takes, and you will find he is worth much more than he costs. Tree farming is a good business but has limited profit margins. You can afford the best professional help obtainable, not the serious mistakes of ignorance or inexperience.

13

Sale of Tree Farms

Tree farms must often be sold to raise cash, to take advantage of price rises that appear to be temporary, or to change the form of assets so that they may be more easily managed. Many principles discussed in Chapter 4 apply to the sale of the entire tree farm, but the sale of land is an art in itself and requires considerable study on your part. Whether you are a buyer or a seller, you will benefit from a thorough knowledge of this aspect of tree farming.

PRICING

Things sell for what they are worth, and the sooner you get the money the better. You must never forget these rules when selling. Although sellers commonly disregard them, they do so at their own expense and invariably encounter costly disappointment.

The greatest mistake you can make is to offer your property for more than it is worth and more than you will sell it for. You may think that this will give you a chance to come back with a counter-offer when you receive an offer from a prospect. You are dead wrong. Prospects make a reconnaissance of the property, recognize the overpricing immediately, and vanish into the night, never to be heard from again. Their interest was whetted by the initial offering but killed by the first investigation; you cannot revive it without a whopping price concession. All you have done by this maneuver is destroy your first impact on the market. Everyone experienced in real estate sales realizes how important it is to capitalize on the interest that is aroused by the first offering. Setting the proper price enables you to do so and accomplish the other main objective at the same time.

The objective is to get the money now, and the importance of doing this can be readily recognized by referring to the example discussed in Chapter 9. That property was worth $60,000 the day it was offered for sale and the day it was finally sold one year later. Failure to set a realistic price cost the lumber company $4800 in interest and about $1000 in ad valorem taxes and management expenses. It is impossible to raise the price of a property on the market to compensate the seller for timber growth that might be taking place or expenses he might be incurring during the sales effort. Furthermore, some expenses are hidden, since the seller does not usually make allowance for the time he spends in negotiations with prospects. These considerations place a high premium on the quick sale.

The first essential step toward this goal is a tree farm inventory. It should be as up-to-date as possible and prepared by a competent, independent authority. Such an inventory enables the seller to set a realistic price and is a powerful tool in selling. Buyers know its utility, and it is effective in arousing their confidence. Exaggerated and unsupported statements about timber volumes and values are quickly recognized for what they are and often create enough suspicion to scare off people who might otherwise be excellent prospects. Tree farms are bought as investments, and the buyer will supply all the romance necessary to consummate the sale.

Chapter 9 discusses an investor's approach to tree farm purchases on the wholesale market, and you should use this approach to set the price in most cases. There is a ready market for properly priced tree farms, and almost no market at inflated prices. Chapter 9 also mentions the role of dreams and pleasure in the retail market for tree farms, and I cannot discuss each possible situation. Your local real estate broker can help you. You must realize which market applies to your property; retail prices are a waste of time in wholesale markets.

MARKETS

The most obvious market is a major company that owns land nearby or adjacent to yours. Such a company is often willing to pay a high price and is easy to deal with once you understand its requirements. Although the company will insist on checking it, your timber inventory will reduce the time required for field examination. The company will be wary of adverse possession and require that it be corrected or that affected portions be excluded from the sale. It will make a thorough search of title and perhaps require that you furnish title insurance or some other satisfactory evidence of ownership. Major-company land purchases must be approved by several departments and echelons of management, and large transactions often require the approval of the board of directors. The company representative will be glad to describe its procedure to you. It will need an option for the time necessary to complete this procedure and will be unwilling to pay more than a nominal amount for it. It will usually make a reconnaissance and will perhaps complete some of the field examination without an option, and it will generally not take an option unless it intends to purchase the property.

The most profitable market is that made up of your neighbors. Every man is naturally interested in buying land that adjoins his and will make a special effort to do so if he can get the money. They represent a chance to sell the property at retail and are often willing to pay a premium because of the location of the property or because they intend to convert at least part of it to another use.

The largest market consists of investors. Price is all-important here, and the only difficulty is reaching them. Newspaper advertisements are effective, and some states publish a farm market bulletin that provides free advertising for the owner of the property. Other tree farmers are obviously interested in this form of investment and may be located through county tax rolls or by diligent inquiry among local foresters, sawmill operators, and pulpwood dealers.

SELLING PROCEDURE

Your offer to sell should be written in a letter or brochure that can be conveniently mailed. It should cover every detail of interest to a purchaser; by listing all the facts (and facts only), your presentation will make it easy for each prospect to see how desirable the property is. A copy of the tree farm inventory and a letter describing terms and conditions are an excellent way to present the property. A thorough written offer to sell will both interest the serious prospect and discourage the idle curious. You want to do both, since sightseers waste valuable time.

The offer must state the price and terms. Cash is always desirable, usually scarce. If the sale is for cash, you should investigate possible sources of credit in order to make constructive suggestions to prospects. If you plan an instalment sale, you should state the terms available to prospects of necessary financial standing.

Every serious prospect will want to make a ground reconnaissance of the property before attempting a detailed appraisal, and you should allow him whatever time is necessary to do this without cost. The amount of free time needed varies widely; you should be cautious in giving it. You must avoid those who profess great interest, ask for and obtain a 30-day option at little cost, and then peddle the property all over the world at an increased price. You must offer the property subject to prior sale and refuse to grant special privileges for longer than a few days to any prospect unless you receive adequate compensation.

You can obtain adequate compensation by selling an option to buy. An option grants an exclusive right to buy on certain conditions for a definite period and should be drawn up with the help of your lawyer. For the protection of both parties, an option should cover all important points in an unmistakable manner. The consideration is a matter for negotiation but is often 0.5% for each 30-day period and may or may not apply on the purchase price.

You can do this also by entering into a purchase and sale contract. The purchaser agrees to buy the property on certain terms and conditions and offers earnest money in an amount sufficient to guarantee performance or compensate the seller for failure. The seller agrees to sell the property on the same terms and conditions and agrees that earnest money will be held by a suitable escrow agent. The amount of earnest money is negotiated also but is usually less than 5% and applies on purchase price. Banks and real estate brokers often serve as escrow agents and will insist that the contract provide definite instructions about how and when to disburse the earnest money. Failure of the buyer to fulfill the contract causes him to lose the earnest money. Failure of the seller to fulfill the contract may cause a lawsuit. Although the contract has a time limit, the time period usually is not fixed as in an option but varies according to the time schedule set up for the different activities of the purchaser and the seller. Here, again, you should consult your lawyer.

METHODS

Nearly every seller tries the do-it-yourself method first, since it has the apparent advantage of eliminating a commission. This advantage, however, is more apparent than real. In selling a tree farm, it takes time to prepare the offering letter, to show the property to prospects, to answer numerous questions that arise, to conduct negotiations before the contract is signed, and to complete closing activities. If your time is valuable, you will usually find that the attempt to save the cost of a commission is false economy. Moreover, the sale of a tree farm is a highly technical business, and there is also a psychological advantage in negotiating through a representative.

For many reasons, use of a sales agent is the best way to sell a tree farm. First, the agent does everything necessary to bring about the sale at his own expense and suffers the loss if he fails to earn his commission by selling the property. Second, an experienced agent creates an atmosphere of urgency that brings buyers to the moment of decision quickly. Third, he is expert at translating favorable decisions into action. A verbal agreement to buy a tree farm is of doubtful value and must be followed quickly by written contracts and considerations. Fourth, an agent knows the intricate details involved in closing a sale and makes an excellent supervisor to expedite these matters. Fifth, an agent is an expert on human nature and senses those characteristics that make your property attractive to each prospect. Quite often he sees many things you do not. A wise old lawyer once said that a lawyer who represents himself has a fool for a client, and the principle involved here usually applies to the sale of tree farms.

You should use great care in selecting an agent; he must be experienced, ethical, and capable in order to represent you properly. As a general rule, real estate agents specialize in properties of a certain kind, and you need a specialist familiar with all details of tree farming. The business is a technical one, and prospective investors want facts, not fiction or high pressure. To interest serious prospects, particularly forest industries, your agent must be able to present these facts from a background of personal experience. Many states require that real estate agents be licensed, and you should investigate this, since the law may penalize you for dealing with an unlicensed agent.

Once you select him, you should deal with all prospects through him alone, even though this seems an unnecessarily complicated procedure at times. If you negotiate directly with prospects, you invariably cause confusion, delay, and unnecessary expense, and often cause failure. Remember that you have retained a professional, and leave the selling job to him. You should insist, however, that neither he nor any cooperating broker ever quotes the property at a price above the listed price. This practice is common and is a sure sign of an amateur. It is done in order to allow price concessions in the future, is never successful, and actually hinders the sale, since it discourages many prospects. You will have no trouble with your agent if you select him carefully, but he may cooperate with agents in other cities who are not personally known to him. You should have a clear understanding about this.

Some tree farms are sold at auction, an effective means under certain circumstances. The property must be attractive enough to interest many people and must be presented in such a fashion that its value can be easily appreciated. A satisfactory auction is a job for a specialist, and you can locate capable auction companies through real estate brokers near you. It is difficult to predict the results of an auction, and you will incur much of the expense of conducting the sale whether the property sells or not. You should exercise great care before adopting this method.

Sale agreements with real estate brokers are usually covered by agency contracts. These are often drawn by filling in the blanks of a printed form, and they cover such details as price, terms, legal description, commission, length of agency, division of option or escrow funds, and so forth. Your lawyer should review it before you sign it; if it is drawn by the broker, it may take better care of him than of you.

EXPENSES

The biggest expense is the commission. It varies from almost a nominal amount to as high as 15%, depending on the size of the transaction, the merchantability of the property, and the services the agent is required to perform. Ten percent is common up to $100,000, and smaller percentages may apply as the size increases. Much of the agent's work is done when an option or sale contract is signed, and the contract reduces his chances to sell the property if it is not exercised or fulfilled. He may also have to do a great amount of work shortly after the contract is signed. Therefore, he is entitled to a much larger percentage of option money or earnest money if the buyer fails to fulfill the contract.

Whether the seller provides a certificate of title, an abstract, or title insurance is a matter of choice, but an offer to sell that includes some provision for showing good title to the property greatly increases its merchantability. Title insurance offers an excellent solution, and your attorney can tell you about its cost and that of other types of title information.

Some deeds may require tax stamps in proportion to the amount of consideration. The seller normally pays for them, but the buyer may agree to do so. They do not have to be affixed to the deed before it is recorded, but they must be affixed eventually. Minerals are often separated from the surface at the time of the sale and retained by the seller. These minerals may then become subject to ad valorem taxes, and many states allow the mineral owner to buy mineral stamps in lieu of all future ad valorem taxes. These stamps must be affixed to the deed and should be paid for by the party who retains the minerals. Recording of the deed is usually the responsibility of the buyer, since it is public notice of his purchase.

PIECEMEAL SALES

You can sometimes get a better price for a tree farm by breaking it into smaller parcels or by dividing it into its components. The number of buyers with $20,000 is much larger than the number with $200,000. Proper subdivision may enable you to interest buyers whose desires vary widely; one part of a tract may be good for a small farm, while another may be ideal for a country retreat. Many timber buyers cannot buy land because of lack of capital, but will pay top prices for timber that can be cut now, so a separate sale of timber will bring them into the market.

When dividing the tract into parcels or components is feasible, you can often hold a sealed-bid sale. You must furnish very detailed information on all parcels and then ask for bids on the timber or the land or both on each parcel. The sealed-bid procedure eliminates the need for setting the right price in advance. Mechanics of holding a successful sealed-bid sale are discussed in Chapter 4.

Unless the tract lends itself to such a sale, the piecemeal approach has disadvantages. First, it is slow and requires much negotiation. Loss of interest on money and the amount of time necessary to deal with many prospects may more than offset any premium prices obtained. Second, the sale of certain parcels or components may reduce the value of the remainder. Separation of minerals, for instance, often seriously reduces the value of the surface. Sale of timber means that you must allow sufficient time to cut and remove it, and there are few buyers for the surface during the period of the timber reservation. Sale of certain parcels may reduce access to the remainder, and the effect of this on prospective buyers is hard to estimate in advance. You should consult an experienced real estate broker and your consulting forester before starting such a program.

TRADING TALK

The real estate market offers a chance to hear statements and questions that illustrate several important facts about the sale of tree farms. They are repeated over and over again, though they amount to little more than nonsense.

"I will not sell for less than $ _____, if I have to keep it forever." This remark is made by a man who does not realize the power of interest. He does not understand that $1000 today is better than $1050 a year from now because ownership of money now enables him to earn interest, whereas ownership of land makes him liable for ad valorem taxes. If economics does not crush the speaker, timber growth may eventually make his land worth the price.

"Would you sell for a lower price?" Nobody should say to a young lady, "If I asked you to marry me, would you?" but the first question is common. The way to get an answer is to make a firm offer and watch results.

"Although your property is an excellent buy at the price, I don't want it now." Assuming the speaker can afford your price, this really means that the property is not an excellent buy, or he would want it. The speaker does not want to hurt your feelings. You should ignore such idle remarks.

"I'll have my offer in next week. Can you give me until then?" This is a request for a free option, and you should reply that the property is offered subject to prior sale and that you hope you will be able to consider his offer when it arrives.

"Several people are anxious to buy my property, but I will let you have it if you will hurry." This obvious attempt to create a sense of urgency usually fails and often means that the speaker has no other prospects. Proper pricing is normally sufficient to create interest among buyers.

"Timberland around here is selling for $ _____ to $ _____ per acre, and there is very little that can be bought. You ought to get at least that much for yours." The speaker is usually repeating gossip current in the business community and probably has no personal knowledge of your property or others that have recently been sold. You should ask him if he will make you a firm offer at these prices.

14
Loans

Every businessman needs credit, and the tree farmer is no exception. Moreover, he has special problems along this line because his needs are unusual. He requires credit in large amounts and for long terms. Long-term credit is harder to get than short-term credit; note the bond quotations in any financial paper. His business is relatively new and untried. Tree farming came into its own as a business within the last 30 years, and many lenders still remember disasters that accompanied timber bonds of 40 to 50 years ago. Nevertheless, money lending is competitive, and the supply of credit for tree farmers is expanding. Let us discuss some sources available now.

FARMERS HOME ADMINISTRATION

The Farmers Home Administration (FmHA), an agency of the federal government, offers real estate loans to family-size farmers to buy, refinance, and develop farms. These loans are made for almost any farm enterprise, including forestry. The present maximum FmHA loan limit is $100,000 with a total indebtedness of $225,000. Since appropriations for this type of loan are limited, an applicant is given priority when 50% of the total loan is provided by another lender. The present interest rate is 5%.

Funds available, loan authorities, and eligibility requirements change constantly, so anyone interested in an FmHA loan should get in touch with the local office. If it is not listed in your telephone directory under U.S. Department of Agriculture, you should write Farmers Home Administration, U.S. Department of Agriculture, Washington, D.C. 20250.

131

FEDERAL LAND BANK

The various Federal Land Banks offer tree farm loans, and the help you can get from them can be explained by describing the situation at one of them. The Federal Land bank of New Orleans makes loans on timberlands through 34 Land Bank Associations, 13 in Alabama, 12 in Mississippi, and 9 in Louisiana. These associations serve every county in the three-state area, and many have branch offices for the convenience of farmers in their territory. Any association can provide applications for loans and detailed information about them.

Any person, partnership, farming or timber-growing corporation (one that derives at least half of its income from these sources) now engaged or soon to be engaged in tree farming is eligible. Owners of properties under lease to industrial companies are eligible if the lease terms are acceptable to the Land Bank.

Eligibility for a loan depends mostly on the ability of a borrower to repay. Several factors work together to establish this: value of the land, value of the timber including rate of growth, and other sources of income available. The landowner must have a sound financial position, a good credit rating, a demonstrated ability to manage a tree farm (either with or without help of a professional), and a sound management and development plan.

Loans are made on stands of both pine and hardwood timber with sufficient volume of merchantable timber. There should be enough annual growth to provide a substantial yield for the term of the loan. Loans are also made on young pine timber smaller than merchantable size, provided the trees will reach merchantable size in less than five years. Proceeds of loans may be used for a wide variety of purposes.

Each tree farm offered as security is appraised to determine its loan value. Land Bank appraisers arrive at loan value by using normal stumpage prices, giving consideration to present volume of timber that may be merchantable under usual conditions. Loan value is also based on the appraised value of the tract for continued production of forest products.

There is now an application fee of $75 with a minimum appraisal fee of $100 plus an additional 50¢ per acre for each acre of woodland in excess of 100 acres (excluding cut-over land or land with little merchantable timber). Where there is a recent, reliable appraisal on the property, the appraisal fee may be less. There is a loan-closing fee in some associations. To prevent misunderstanding, applicants should call the local Land Bank Association to determine the total amount of fees.

The amount of a loan depends on the appraised value of the land and timber. It usually varies from 50 to 85% of the appraised value of the property at today's prices. Loans to any individual or entity may exceed $24,000,000, if financial records, repayment ability, and so on, are satisfactory. The Federal Land Bank of New Orleans can approve loans up to $1,000,000, and most associations can approve loans up to $300,000. Larger loans must be approved by the Farm Credit Administration. Since passage of the Farm Credit Act of 1971, more authority has been delegated to the Land Bank, which in turn has delegated more authority to the associations. This has speeded up closing of loans and handling of releases. The average time for closing loans is now 53 days; many are closed within 30 days.

Terms of the loans range from 5 to 40 years. The principal can be repaid quarterly, semiannually, or annually, and it can be paid in full at any time without penalty. Principal payments can be deferred during periods of development of timber; you might be able to pay interest each year and make principal payments every five years.

All Land Bank loans are made at a variable interest rate, which means that the rate goes up or down with the cost of money. Wide swings in the cost of money in the recent past caused this change. The rate in mid-1977 was 8¼%, reduced from 8½% on March 1, 1977 because of a favorable money market. No federal money is involved in the Land Bank system; the Bank secures most of its funds from the sale of bonds to the investing public.

Since the Federal Land Bank is owned by its members, all borrowers purchase 5% of the loan in stock. When the loan is paid in full, the stock is retired at face value. If earnings permit, a dividend is paid on the stock. Because the money market situation has been so unfavorable, no dividends have been paid for several years. At the outset, this increases the effective interest rate to 8.68% and to higher rates as the principal is repaid.

Annual growth or its accumulation, as established by the appraisal, may be released without payment on the principal of the loan. For release of more timber, principal payments in proportion to the value will be required, although there may be exceptions to this rule.

The Federal Land Bank has some advantages for borrowers. It has been a dependable source of credit for more than 60 years. Its variable interest rate assures a rate consistent with the cost of money. There are no repayment penalties. Credit life insurance is available if the borrower is in good health and under 66 years old.

In August 1976, the Federal Land Bank of New Orleans started a rural appraisal service available to farmers and tree farmers with headquarters in Jackson, Mississippi. The service offers three types of appraisals:

1. *Letter appraisal.* This brief report gives the estimated value of the property and is not intended as a basis for a credit decision by the Federal Land Bank.

2. *Letter form appraisal.* This more detailed report is intended for use by the Federal Land Bank or other lenders as a basis for a loan and is made on forms approved for that purpose.
3. *Comprehensive appraisal report.* This detailed report includes information on soil classes, uses of property, building values, and so on. It may also be used by the Federal Land Bank or other lenders as a basis for a loan.

The Land Bank charges fees for these appraisals, depending on the type needed. Its appraisers offer no technical advice or assistance, but merely evaluate the property. They are often trained foresters and always have experience in timber appraisals.

COMMERCIAL BANKS

Commercial banks also lend money on tree farms. Regulations governing these loans are issued by the Federal Reserve Board and the comptroller of the currency, and each bank has a copy of the latest regulations. As a general rule, they can lend 40% of the value of merchantable timber for a maximum of 10 years. This term is acceptable only if 10% of the principal is amortized annually; otherwise the maximum term is two years. Since the value of most properties includes some consideration for land and reproduction, this regulation reduces the utility of commercial bank credit.

Additional funds may be available from commercial banks to borrowers who have already established satisfactory financial relationships with them. Essentially, however, banks are sources of short-term credit with the terms of loans measured more in months than in years. They depart from this practice only with reluctance and for good reasons and, therefore, are not active in the tree farm loan field.

INSURANCE COMPANIES

The most active and satisfactory lenders are insurance companies. Most larger companies have some experience in the field, and the extent of their interest in loans of this kind is increasing. At least one company has a graduate forester as a tree farm loan manager, and all have access to competent professional help.

These loans carry an interest rate of about 9% at present, and, although the maximum term is often 20 years, later refinancing is easy if experience has been satisfactory. Principal repayments may be postponed for the first few years. Prepayment privileges vary by company; one allows prepayment without penalty, provided that not more than 20% of the principal is retired in any one year. As a general rule, insurance companies are not interested in loans of less than $50,000.

Insurance companies are restricted by law to loans of less than 75% of appraised value or market value, whichever is less, and they prefer to restrict loans to 65% or under. This is substantial help, however, since the investor can borrow as much as $3 for every $1 of equity capital he contributes. The companies require financial statements from the borrower, showing that his net worth has the proper relationship to the amount of the loan. They expect the borrower to show some financial ability, since they do not want to acquire property by foreclosure. Few tree farm loans stand on their own bottoms, and the lender often requires personal endorsement of the note.

Mortgage-loan departments are organized in echelons, each echelon being limited in the amount of loan on which it can give a definite commitment. Most tree farm loans are so large that they require extensive appraisal work by lower echelons and final approval at the home office. Therefore, most companies require a standby fee of approximately 1% with the application. If the application is approved and the loan is not accepted by the applicant, he forfeits the standby fee to compensate the company for work done by its employees. Otherwise, the fee is refunded.

Appraisals include some value for merchantable timber, reproduction, land, minerals, and other features producing income, and companies usually require a recent timber inventory by a qualified authority, even though the amount of the loan is limited by the price of the property. Here again the company must assure itself that the value of the property is sufficient to retire the interest and principal of the loan. Some companies require that the timber inventory be quite detailed, and you should investigate these requirements if you plan an inventory to be used for loan purposes. Appraisal values for merchantable timber are close to current market values. Occasionally, an investor uses a loan to purchase a property covered by a cutting contract providing for specific annual payments in cash. If the party agreeing to make payments is obviously well able to fulfill this contract, the lender may not require a timber inventory but will require its employees to make a thorough inspection of the property on the ground. Contracts of this kind make obtaining a loan much easier in most cases. Time required from date of application to date of final loan commitment is from 30 to 60 days when the amount is near $1 million, assuming full and intelligent cooperation from the applicant, and faster action is possible on smaller amounts.

Some companies obtain loans through exclusive brokers, and others deal directly with the borrower. Although either method is satisfactory from the borrower's standpoint, you should ask for a detailed statement of charges connected with obtaining the loan. During preliminary stages, you should shop around for the best terms, but you will have to pick the company before you file an application. Most require title insurance on the property to be mortgaged. Since all companies are subject to the same fluctuation in the money market, you will find almost no difference in the rate quoted by each company. Borrowers with exceptional financial resources may be able to shave the rate as much as 0.5%. These companies are experienced and competitive, and they can tailor a loan to fit almost any conditions. Mortgage-loan managers continually study businesses of all kinds and can give you valuable advice.

You will notice that insurance companies and the Federal Land Bank use component-parts appraisals in establishing the loan value of a tree farm, although I said earlier that this practice is a mistake unless great care is used in assigning values to various parts. They correct any mistakes that might be made by requiring the borrower to show how the loan will be repaid. This means that they are concerned mainly with the return the property will produce.

INDIVIDUALS

Individual capitalists provide another source of tree farm loans. Although these individuals are not formally organized and must be searched out, they can be found in every town in a timbered section. They seldom have capital for loans of over $50,000, but long-term credit in small amounts is not supplied by any other source. They are often familiar with the value of tree farms near them and are anxious to obtain a fixed return of about 8%. Most are not unwilling to acquire a tree farm by foreclosure. They may be willing to lend 90% of the purchase price for a term as long as 20 years, particularly if the loan is amortized by a series of regular monthly payments.

These individuals form a source of great potential help to the business of tree farming everywhere. They can be especially helpful to foresters who desire to acquire a small tree farm for themselves. Such individuals may be located by diligent inquiry among banks, lawyers, and certified public accountants of the community.

SELLERS

Credit in large amounts is provided by owners of tree farms who are willing to sell them on terms. Potential purchasers should explore this possibility. A seller sells something because he needs money, and he may be willing to make concessions to get it. Since many sellers are individuals and can seldom obtain a return on capital more than the interest paid on savings accounts, they will often agree to an interest rate of about 8%. They usually think in terms of much less than 10 years but may be willing to postpone principal payments until late in the life of the loan.

Sellers also extend credit to take advantage of federal tax laws. Many have a substantial profit in the property to be sold and are anxious to spread the tax on this profit over many years. In general, a seller can spread his profit, and therefore his tax, over several years if he receives less than 30% of the sale price in each year. This provision works to the advantage of both buyer and seller. Changes in tax laws are made constantly, however, and you should consult your tax adviser when considering such a trade.

SUMMARY

These are the present sources of tree farm loans. Although there appear to be adequate funds for this purpose, tree farm loans are not as common as might be expected. One reason for this is the low return earned by many tree farms; loans are unsatisfactory for both parties when the enterprise earns *less* than the interest rate. Tree farmers can remove this limitation by increasing yields and profits. Another reason is that there are few sources of loans less than $25,000, since small loans are costly to negotiate and service. A solution to this problem may arise as the business grows.

Some charitable foundations are moving into the field, and many provide quite flexible loans. Investment policies of these foundations are often solely in the hands of their directors or trustees, and they may be willing to provide funds on a combination of equity and debt that permits low interest rates in anticipation of large future profits. Some state governments offer loans to certain classes, particularly veterans, on lenient terms. Real estate brokers and consulting foresters can help you with information on this.

If none of these agencies meets your needs, please remember that tree farming is a relatively new business. A large volume of credit is available to a business only after long and satisfactory experience. All lenders are watching your progress with keen interest; you can help yourself and all other tree farmers by succeeding.

15

The Large Industrial
Tree Farmer

Although their financial position and objectives are much different, large
industrial tree farmers are of great interest to you. Often, they have been
tree farmers for any years and can concentrate great skill, energy, and brainpower on problems of the business. Some things they learn can be applied
directly to your tree farm, and an insight into the workings of their corporate
minds may help you predict the future. It will certainly help you in appraising tree farms as investments and in trading with these larger entrepreneurs
for land or timber. They have hundreds of millions of dollars invested in tree
farms and work constantly to improve their operating results. I can discuss
only a few points now, but I urge you to study these companies continually.

ATTITUDE TOWARD LANDOWNERSHIP

A corporation is an individual created by law, and it is just as much an
individual as you are, in terms of having a limited amount of money. Although this is hard to visualize when you see one with thousands of
employees, several manufacturing plants, and hundreds of thousands of tree
farm acres, just imagine what will happen to this great amount of capital if
every deal the corporation makes is a bad one. Even the largest will be
struck down almost overnight if it violates the laws of economics. This means
that it must watch every penny of expense and make every dollar of capital
work at maximum efficiency. A large company has a much harder time at this
than you do, because its employees do not have direct and immediate contact with profits and losses, and the board of directors is often the only group
that fully realizes how limited company funds are. Let us study the problems
of the XYZ Paper Company just before World War II.

XYZ has $100 million invested in a paper mill, warehouses, and so on, and its main business is the manufacture and sale of paper. If for any reason it cannot make paper, these installations soon become worthless except for what can be salvaged from junk dealers. Until an adequate substitute comes along, wood is essential as a raw material, and the directors are constantly haunted by fears that it will become unavailable or rise in price so fast that any profit will be impossible. They feel that they must provide some insurance against such a catastrophe, and they can do so by leasing or buying land on which to grow this wood. If their mill were the only one in the country and there were millions of acres of trees waiting to be harvested, they would not spend a nickel for tree farms, but this is hardly the case in the United States. Our population is growing rapidly, and tree farms are disappearing into other uses. XYZ also feels the hot breath of competition on its neck and fears that it will be beaten out in the race for the raw material available. Therefore, the directors are faced with a formidable decision—is tree farm ownership necessary?

To help them make this decision, they investigate every conceivable aspect of the problem. How much wood is being grown within economical hauling distance of the XYZ mill? Who is growing it, and how much of it is likely to be available to them? How many other companies of all kinds are using wood in the area? How much timberland is controlled by these companies for their own use? Are there any national forests in the area, and what is the condition of these stands? And, most important of all, what about individual tree farmers like you? Do you own much land? Are you doing a good job of keeping it in the most productive condition? *Are you making money at tree farming?* If you are, their decision is much easier, since you will provide most or all of the necessary wood, and the profit you make will keep you in business. Your state of mind and profit and loss statement are extremely important to them, since you own most of the land in the country and have more investment capital than all forest-products industries combined. Therefore, they will watch for changes in your attitude that might be disastrous for them. After considering everything, they decide that some tree farms are essential, and they now must answer another question—how much land should they buy?

The most comfortable position would be to be able to grow all the wood they need, but this may not be a satisfactory solution, even if they have, or can get, the necessary money. Stockholders own the company, elect the directors to run it, and demand that it pay them dividends and also increase in value over the years. What about return on stockholders' capital? If this falls too low, the stockholders would be better off to liquidate the company and put the money in savings accounts. Furthermore, the company's life depends on its profit over the long term. If the ABC Paper Company earns a better return, it will have more money to expand and improve its operations and eventually acquire many of XYZ's customers. XYZ's bankers do not like shrinking profit margins either. If the bank rate is 8%, XYZ must use the proceeds of any loan to earn more than 8%, or the loan is in danger. Therefore, the directors must reach a compromise solution, and no convenient handbook spells out the details.

If they cannot grow all of their requirements, how much land will grow a certain percentage of present and anticipated future requirements? How much will the land cost? How much is necessary to put it into maximum production? Is it cheaper to spend money on development or on additional land? When is the company likely to need wood from its own land, and will it be available then? Are tree farms of the area being converted to other uses? How is the competition reacting to the situation? The directors must know both present and future answers to these questions, since it takes many years to grow a tree, and since the company requirements are likely to change. These are hard questions to answer in a definite manner, and margin for error seems very small.

There still remains another question—what returns are available on capital invested in tree farms? The directors instruct the accounting department to investigate this and to devise a system of records to show how tree farms will perform if they are acquired. The accounting department encounters trouble immediately, because of the nature of the business and because many foresters understand little about accounting. In addition, all tree farmers have the problem of plus factors that cannot be measured. No set of records will reveal these, and perhaps a good system for all purposes will never be devised. Nevertheless, the accountants report that tree farms will produce an annual return of 6 to 8% on their market value. Market value, not book value, is the proper standard against which to measure the return, for the company must pay market prices to acquire land and, at some later date, will be able to sell all or part of its land. In standard accounting practice for tree farms, book value often has no relation to market value.

This small return disappoints the directors, especially when they compare it to the rate earned by machinery, working capital, buildings, and other parts of the capital. Equity capital is hard to obtain and must work to maximum advantage. They are concerned, however, with the long-term health of the entire coporation. The company's assets must work together; each must earn the maximum possible return but must do so in a manner that permits maximum return on the entire capital over a long period. This consideration leads to the purchase of tree farms; they are necessary to protect the remainder of the capital. Although every dollar invested in tree farms *decreases* the average rate of return on all invested capital, the directors feel forced to buy them for protection. Many large companies have been pictured as land-hungry monsters who gobble up every available acre and keep it for greed alone; nothing is farther from the truth. Almost every board of directors would sell the company's tree farms tomorrow if they were not considered indispensable for the long-term health of the company. These decisions are examples of the awesome responsibilities directors face; the important thing for you is the reluctance with which they invest in tree farms.

What does all this mean to you? It means that companies like XYZ will always do what they can to make you successful. They will buy your timber whenever competitive conditions permit, and tree farming without markets is impossible. They will pass along to you everything they learn about how to increase the production of every acre. Because of the low returns described above, they now conduct extensive research on the biology of tree farming, for example, on genetics and forest insects and diseases, and good business compels them to tell you what they find. They will help you with fire protection, dealing with trespass, boundary-line maintenance, and any other projects that can be reasonably justified. They want you to succeed almost as much as you do.

It means also that they cannot always be counted on to buy your tree farm when you get ready to sell it. They are struggling to keep tree farm investment at a minimum. If you are already a tree farmer, you may have to sell your tree farm in competition with all other investments available on the world market, and this is where returns are all-important. Therefore, land producing at much less than its capacity is clearly a liability, and now is the time to get busy with programs to increase annual production and profit. If you are not a tree farmer but plan to be one, you may now encounter greater opportunities through decreased competition from large companies.

ATTITUDE TOWARD FORESTRY DEPARTMENTS

Let us continue the hypothetical history of the XYZ Paper Company. Since the directors regard company ownership of tree farms as a necessary evil, you can understand some of their attitudes toward forestry departments. When the directors decided to acquire tree farms, they were forced to hire someone who knew something about this entirely new business, and the first forester came on the payroll. He knew little about business in general, and almost nothing about the paper business in particular, but he was hardworking, conscientious, and an expert in biology. He buckled down to a job of monumental proportions and gradually acquired other foresters to help.

One early problem was that neither directors nor forester knew much about the others' business or problems. The forester had to report to someone, and the directors picked the vice-president in charge of operations at the mill around which the first land was bought. This man was already "busier than a one-armed paper hanger with the seven-year itch," knew nothing about forestry, and (when he had time to think about his new responsibility) hoped that the forester would come up with some attractive proposals and programs that he and the directors could understand. The forester, on the other hand, was a novice in the business world and hoped for guidance from above in order to determine how tree farms were expected to contribute to company operations. When no help was forthcoming, he was forced to guess, learn by trial and error, or talk to other foresters who were similarly confused by this new situation. Consequently, the early years were marked by failure, frustration, wasted motion, and poor financial results. Nevertheless, this was a marriage of necessity, and it endured.

Not long after World War II, the great rise in stumpage and land prices began. Every day trees and tree farms were worth a little more, and timber growth made some timber sales possible. During one or two years at about the same time, the paper business fell off, and income from timber sales helped compensate for the dip in earnings from manufacturing. The chief forester began to make real money; stockholders were pleased; and the directors allocated more funds to tree farm operations. Mutual understanding between directors and foresters grew, and this led to increased efficiency. Nevertheless, the chief forester failed to realize that much of the increased profit was due to a great national trend and not to his own capabilities, and he added foresters and adopted programs that were not essential. As long as money was rolling in, the directors did not give their entire attention to a subsidiary operation. In the late 1960s and early 1970s, however, the company found it difficult to earn a suitable return on assets, and when stumpage prices fell as much as 50%, the foresters could not contribute much to profit. When this happened, the directors naturally turned critical eyes toward the tree farms and asked questions.

WHY IS THE RETURN SO LOW?

One reason for low returns is the high cost of management; almost everything XYZ does can be done cheaper by an individual. The salary level of its foresters is not unusually high, but there is a sizable package of fringe benefits. It is hard to give these foresters a sense of urgency about costs, since they are several steps removed from the actual money-making process. They do not compete against anybody, except indirectly, and it is hard to measure the efficiency of their operations. XYZ is the constant target of nuisance lawsuits, restrictive legislation, ad valorem tax discrimination, union organizers, and empire-building bureaucrats within its own ranks. Although these disadvantages are common to all large-scale enterprises, the nature of tree farming does not permit some offsetting advantages that usually accompany large size. Unless ownership is consolidated, the unit cost of management does not fall as the total area increases. The need for brainpower to direct manpower poses a perplexing problem for the company. Until total land area becomes quite large, it seems hard to justify the cost of a capable man who does nothing but think about problems and policies. Since the foresters are almost entirely preoccupied with day-to-day administration and training, this kind of brainpower has been necessary since the beginning.

Another reason is the inflexibility of the operation. XYZ owns tree farms primarily to produce raw material for its mills and must channel the productive capacity of each acre toward this product insofar as possible. XYZ must grow pulpwood and allow trees to be used for other purposes only when it appears unlikely that they will be needed at the mill. Of course, it is possible to make money by growing pulpwood, and this may be the best crop on many acres. The company also finds that it will always grow some sawtimber, and it strives, within limits, to grow the most remunerative products. At all times, however, it must be ready to grow its own raw material, and emphasis must be on management for one product. You have already seen that a tree farmer needs every possible chance for profit, and greatest success depends on complete freedom of management.

Another type of inflexibility makes it almost impossible to sell certain tracts of land for conversion to higher uses. Sales of this kind are very profitable for many tree farmers, but the administrative difficulties of selling any XYZ land are numerous. Every sale must be reviewed by the district forester, chief forester, operating vice-president, financial vice-president, and president at the home office, and what began as a handsome profit is soon entirely consumed by administration.

To a certain extent, federal tax rates affect the efficiency of a corporate tree farmer. All tree farmers, corporate or individual, perform TSI work, a deductible expense, to increase future income. A corporation, with an income tax rate fixed at about 50% regardless of the amount of income, spends $1 of its own money and $1 that would otherwise go for taxes when it performs TSI work. An individual in the 67% tax bracket spends $1 of his money and $2 of tax money. Therefore, such cultural work is a more tempting investment to an individual, in some cases, than it is to a corporation.

WHAT CAN BE DONE TO IMPROVE THE RETURN?

This question concerns both large companies and individuals but has one aspect important only to companies. This is management of foresters, as opposed to forests, and larger companies have naturally assumed the lead in this study. They seek the answer to several questions.

What is the proper workload for a forester? At one time it was commonly believed that a forester would be fully employed with the wood activities necessary to manage 25,000 acres of tree farms, and many organizations were built on this belief. The forester was expected to mark timber, maintain boundary lines, fight fire, plant trees, perform TSI work, and do whatever else was necessary in the woods. Experience has demonstrated that this scheme is expensive and unsatisfactory. Unless he performs common labor much of the time, the forester cannot stay busy under such a program. Therefore, he is idle part of the time, and he leaves the company for another job where his talents and experience are more fully utilized or turns to "moonlighting" out of boredom and the need for extra money. Widespread moonlighting in a company almost surely means that it is overstaffed. The company suffers in such a situation. It loses its good men after it has paid for training them, and it pays the hidden cost of divided loyalty that is a part of moonlighting. This situation cannot be endured. The usual solution is discharge of unnecessary foresters and increased responsibilities and salaries for those retained.

How does this work? Usually very well. The forester becomes a manager instead of a technician. In addition to managing 25,000 acres of tree farms, he buys perhaps 25,000 cords of pulpwood each year, handles all public relations in his area, takes an active part in the civic life of his community, and carries out all the administrative duties he might have if he owned the tree farm as an individual. He takes a solid part in the entire operations of the company. In addition to doing forestry work, he gets wood orders from the wood-procurement manager. He prepares financial forecasts of income and expenses for use by the accounting department. His land-acquisition work, if any, brings him into close contact with the company's attorneys. He may have several skilled laborers on his staff or have access to them from a company labor pool, but he continually looks for the chance to get work done cheaper through the use of contractors. He becomes what amounts to a small businessman and may be one of the busiest and happiest in the community. The value of the assets he manages may exceed $2 million, which is substantial responsibility.

What does this do for the company? First, it produces happy employees because they are busy and well paid. Second, it allows the company to obtain and keep the best men available. Its employees are the envy of their colleagues. Third, and perhaps more important, it allows the company to build a broad base from which to select future top management. A company with millions or hundreds of millions of dollars invested in tree farms needs management of the highest order, and men capable of this must be trained for many years. One large company has devoted great thought to the management of foresters, using outside management consultants and its own executives. The system it developed has been in use for over 20 years and seems to be producing a dedicated and efficient forestry staff. The low return on company tree farms makes this essential.

What does this mean to you? Major-company experience in the management of foresters may help you decide whether your workload is sufficient to justify a full-time forester or whether your present employees are utilized to maximum advantage. Most companies are happy to share this experience with you and can save you the expense of trial and error. In addition, all tree farmers benefit from talking shop with others in the business, and the information you obtain from a forester with great and varied responsibilities is apt to be more valuable than that obtained from one who deals with a limited sector.

HOW THE TREE FARMER CAN HELP HIS CUSTOMER

Another reason for studying big companies and their problems is to enable you to lend them a helping hand from time to time. They do not expect you to fight their battles, but they sometimes need your help to obtain fair treatment. Whether you realize it or not, you and they are partners because they buy what you sell, and anything that imperils a good customer jeopardizes the business of the seller. There are times when you, as an individual, have more power than a big company.

Take, for example, discrimination against absentee landowners by the county government in assessments for ad valorem tax purposes. Higher taxes make tree farming less profitable and may force sale of the land if carried to extremes. Companies buy tree farms only because they are necessary for the long-term health of the enterprise; reducing or eliminating the small return that is available means disposal of the land and cessation of operations. It is not necessary for the county actually to carry out liquidation by taxation. All it has to do is threaten to do so by continual tax increases and discrimination. Companies can read handwriting on the wall at great distances and will gradually pull out of governmental jurisdictions where taxes are unfair or unreasonably high. You exert some control over your government by voting, and companies have no vote at all unless they have employees living in the county. What you say to elected officials has urgent meaning; what the company says is of academic interest. Therefore, do not hesitate to speak up in opposition to unfair treatment. You speak in your own behalf, even though the company may be the immediate beneficiary.

BIG COMPANIES AND THE MARKETS THEY PROVIDE

The foregoing discussion of problems faced by the board of directors indicates some factors that influence the raw-material-procurement policies of the XYZ Paper Company, and these policies, both present and future, mean a lot to you. XYZ must ride both ends of a seesaw. It needs to buy wood as cheaply as possible to compete with other paper companies, but it must pay as much as possible to make tree farming profitable for you and keep you in business. It further recognizes that the cost of harvesting or producing wood is a major factor. Labor does not move readily into the woods, so men working there now must be able to earn enough to live adequately. Most wood is produced by small, independent contractors who must be able to meet payments on trucks and other equipment in addition to earning a living. If the delivered price of wood is not sufficient to satisfy both you and the producers, XYZ must buy enough land to grow its entire requirements and must hire enough men and equipment to harvest it. Such a project is impossible from the standpoint of capital requirements alone.

As far as XYZ is concerned, the cost of wood delivered to its mill is the important thing, and this depends largely on transportation costs. XYZ is on another seesaw here. The cheapest wood is usually that trucked directly to the mill yard, and in certain situations and at certain times each year XYZ can operate its mill on wood purchases at the yard only. Nevertheless, to depend entirely on this supply is dangerous. What happens during periods of local heavy rains that prohibit woods operations? What happens when production labor moves out of the woods for seasonal agricultural work? These two possibilities (and others) make the company spread its wood-procurement territory over a wider area than cost alone dictates. It buys its wood through either company woodyards or dealers located along railroads leading to the mill, and, since some capital investment is required, it chooses buying points with great care to ensure permanent operation. The wood-procurement manager then maintains constant, almost daily, contact with someone at each buying point to keep the amount of wood on hand at proper levels and to ensure its orderly arrival at the mill. Having to stop the mill for lack of wood is a calamity to be avoided at almost any cost. On the other hand, an excessive inventory is a financial burden, and some of it may be lost through insect attack and other deterioration. Therefore, the life of the wood-procurement manager is full of nightmares.

He has other problems of a long-range nature. When the company expands, he must procure larger quantities and must know whether they can be obtained and how much they will cost. If the species used by the mill are limited, he finds tree farmers unwilling to permit cutting of only one species, or a small number of species, and this makes his job harder and the wood more expensive. Therefore, he is constantly urging the research department to develop processes for additional species. In addition, there is always the chance that the entire raw-material requirement will change, for instance, in a complete switch from hardwood to pine, and he must be ready to get whatever the mill needs.

What does all this mean to you? First, it means that the prices paid by buyers are set by factors much more closely related to the national economy than to the desires of an individual company. The cost of growing timber and the cost of production labor and equipment necessary to get it to market are nearly equal throughout a wide area, and this explains why the delivered price of pulpwood is the same for a long distance along the railroad. The wood-procurement manager cannot achieve many economies here. Therefore, any price increases on wood in your area will come from changes in the whole economy of the paper business, not from negotiation with an individual company. Second, it means that the location of your property or prospective purchases with respect to established buying points, and the distance of these points from the mill, is very important. The high cost of transportation forces the wood-procurement manager to make every effort to buy wood close to the mill. He cannot cut the delivered price at buying points; he can only minimize or discontinue purchases at distant points. Third, it means that fleeting market opportunities may be available to you. In spite of the most strenuous efforts, wood-procurement managers cannot anticipate everything, and it is sometimes necessary to buy wood at great distances from the mill. Such buying may continue for only a few weeks, and you should be alert for this possibility if your tree farm is located in what is called a "fringe area." Fourth, it means that some trees that cannot be sold now may be in brisk demand soon. Even large mills must convert to other products when their markets are usurped by competitors, and all buyers are aggressively seeking uses for species that are unwanted and therefore cheap.

CONCLUSION

To simplify the discussion in this chapter, I have used the example of a paper company, because it is usually a large enterprise with fully developed problems. The principles involved apply to all large industrial tree farmers.

It is important to understand major companies as thoroughly as possible. Their resources are great and apt to be used in a small area where their effects are greatest. If, for example, a large paper company becomes concerned about the dwindling supply of pulpwood in the area within 35 miles of its mill and decides to acquire an additional 100,000 or 200,000 acres of tree farms, prices are apt to rise until the program is complete. An expansion of its capacity or operating ratio or a change of species needed will greatly alter the outlook for tree farmers nearby. Therefore, those large companies operating near you deserve your careful and continued study. Correct judgments about their future actions may bring you large profits.

16

Guesses About the Future

Foresight is priceless, especially in tree farming, which is a long-term business that requires thinking far into the future. Timber growth on a specific property can often be predicted accurately, but trends of a more general nature are also important. One major attraction of tree farms as investments is the steady rise in the price of all land; everybody knows this, but few can predict it accurately. Anything that helps you with this prediction contributes to your success as an investor. Although I am no seer, I think that some trends are already evident and will work great changes in the future.

NO SHORTAGE OF TIMBER

I predict that there will be no shortage of timber. We have heard cries of timber famine for 50 years and yet have much more than we did in the 1930s. I think those who still cry famine are wrong for several reasons.

First, they underestimate the private tree farmer because they do not understand him. When they see a tree farm producing at only half its apparent capacity, they assume that the tree farmer will not invest in growing more trees because he does not have enough capital or is unwilling to invest in a long-term venture. Almost as a kneejerk reaction, the famine-criers call for more government subsidies.

The tree farmer is a lot smarter than they are. He can calculate return on investment as well as anybody. The reason he has not invested to increase production in the past is that prospective returns did not match obvious risks. (If all tree farmers had planted trees 40 years ago, as they were then urged to do by many foresters, we would have a timber glut of huge proportions, and you can imagine what the prices would be.) If you could calculate the efficiency of his investments, I think you would find him the equal of other asset managers. By following his own wisdom, he has made a profit and produced all the timber we can use and some to spare.

Second, the scarcitymongers must make a very long-range prediction. Since there is certainly plenty of timber for the time being, we will not be in any danger until 30 years from now, and as I pointed out in Chapter 9, most investors know that predictions of conditions even 20 years off are almost worthless. A recent editorial in *FORBES* magazine put it another way: "Anyone who says businessmen deal in facts not fiction has never read old five-year projections." All predictions of timber famine for the last 50 years, and there have been many, have been wrong. Are present predictions better than past ones? I do not think so.

Third, they underestimate the impact of scientific developments. The work of two men, Drs. Peter Koch and Bruce Zobel, illustrate what can be done. Koch has spearheaded development of a machine that harvests not only the whole tree but also much of its stump and central root system. The process appears likely to increase fiber yield from most tracts by 20%. Zobel leads a large force working on tree genetics. His progress so far appears certain to increase tree growth by 20% and perhaps even by 50%. The discoveries of these men and their associates are already part of our knowledge. All that is necessary is to apply that knowledge. Without doubt more discoveries lie ahead.

NO SHORTAGE OF PREDICTIONS OF SHORTAGES

I predict that there will continue to be predictions of timber shortages. Capable bureaucrats have a vested interest in the subsidy programs they helped to set up in the first place and have been administering for years. Furthermore, some of them really believe they are saving us from famine.

Others justify subsidy programs to keep the price of timber products from rising and hurting the consumer. What they would do is take tax money away from you and give it to somebody else for planting trees so that the trees you now own will be worth less in he future.

If tree farming is as good a business as I have said it is in this book, it does not need a subsidy. If the subsidy boosters really want to improve the economic climate for tree farmers, they would urge tax reductions so that tree farmers could spend the money more wisely than the subsidy boosters.

INCREASING IMPORTANCE OF THE INDIVIDUAL TREE FARMER

I predict that the individual tree farmer will become much more important. Although he owns only 60% of the commercial forest land in the United States, he is almost the only owner directly responsive to the market, and by market I mean all demands of all citizens of this country and its world trading partners. Industrial ownerships are managed to produce the products most wanted by the individual manufacturer. Public ownerships are more and more managed to suit the desires of small but vociferous pressure groups who often want only a narrow range of benefits. Both industrial and public ownerships, therefore, lack flexibility and cannot respond quickly to change. This may be a good thing, and if I were the managers of these ownerships, I would probably do what they are doing. But I am certain that the wisdom of all these managers can never match the collective wisdom of the people as expressed in their market actions. Since the individual tree farmer pays close attention to this market and does his best to satisfy its demands, he is the one who, in both the short and long run, will provide what the American people want.

He will become more important because there is coming into being a body of consulting foresters organized to serve his needs. For years he has needed someone on his side; now he is getting them. You cannot expect a tree farmer to commit large sums to investments when the only help he can get with them comes from government foresters with built-in restrictions on their competence and range of services or from industrial foresters with a built-in conflict of interest because their companies are timber buyers. Now that the number of competent consulting foresters is growing every day, the individual tree farmer will make more money from his lands and will increase his investment in them.

He will become more important because there is being developed a means for small, direct investment in tree farms with professional management, a sort of tree farm mutual fund. Although the only vehicle I know about is still in registration with the Securities and Exchange Commission and, therefore, has no performance record, I believe that it and others like it will eventually succeed because tree farms are excellent investments, especially during inflation. They rise in value in proportion to the decline in the dollar; they produce cash incomes from time to time if needed; they need little attention each year (ordinary farms will deteriorate if not regularly cultivated); they are quite liquid so far as the timber is concerned. When the proper investment vehicle is developed and the public becomes familiar with it, large additional amounts of capital will flow into tree farms.

The increased role of the individual tree farmer will be a good thing in many ways. As he has already demonstrated, he will grow whatever timber is needed but not an overabundance. He examines every expenditure and constantly appraises the entire investment. Since he is spending his own money, he makes his dollars work hard and efficiently. His independence guarantees flexibility; he can change forest management plans in minutes to take advantage of fleeting market opportunities or special tax situations. As

Ludwig Erhard of Germany said, "Turn the people and the money loose, and they will make the country strong."

CONTINUED RISE OF THE SOUTH AS THE NATION'S TREE FARMER

I predict that the South, already important as the country's wood supplier, will become much more so. It is undoubtedly one of the most profitable tree farming regions of the world. It has adequate to abundant rainfall, a prime requisite for tree growth. It has fast-growing species suitable for many purposes and easy to regenerate. It has terrain that permits cheap movement of trees to market. It manufactures over 60% of the nation's paper and provides excellent markets for pulpwood, and its markets for other tree farm products are well developed. It is relatively close to large population centers where forest products are used in great quantities. Its foresters have a generation of experience in tree farming and stand ready to apply their experience to meet future challenges. These advantages decrease costs and increase profits and will speed the flow of capital into southern tree farms.

What does this mean to you? I think it grounds for almost unbridled optimism if you own a southern tree farm. Closer attention to the financial aspects of tree farming must result in increased appreciation of the advantages of the South with corresponding interest in its tree farms by investors. Increases in available raw material, relatively easy to come by in the South, bring increased industry, markets, and profits. I believe that future southern markets will be almost unbelievably good by today's standards. Bold and speedy action to put your southern tree farm into maximum production appears likely to pay handsome dividends soon.

EXPANSION IN MARKETS

I predict a steady, and sometimes spectacular, improvement in markets for tree farm products. The total area of tree farms in the United States is decreasing. Two hundred million people need room to live, highways to travel, farms to cultivate, utilities for service, water to drink. Interstate-highway ROW's require about 45 acres per mile, and much of this land comes from tree farms. Uneconomical small farms in timbered sections have nearly all been abandoned and have reverted to timber. Where farming is a profitable venture, however, land clearing is proceeding at a rapid pace. Since this conversion of timberland is based on a sound profit potential, it is apt to be permanent. The tremendous growth of this country creates an insatiable demand for pipelines and power lines, and much of this area comes from tree farms. One large tree farmer has 65,000 acres of land in ROW's alone. Every major city demands adequate water supplies, and a reservoir covering 25,000 acres is becoming a common sight. Water is essential, but flooding of such great areas takes its toll, since the flooded land is often the most productive tree-growing part of the area.

Even if the demand for wood remains fixed, each tree farmer benefits when land is removed from timber production. Mills must run on something, and the supply decreases. Unfortunately, this process does not move smoothly. When the wood supply of a locality shrinks, buyers bid up prices, thereby reduce their profits, and finally cease operations. Markets shrink, and tree farmers become discouraged. This is the time for the courage to make big profits. Many investors have made fortunes in the stock market by selling when everybody was buying and buying when everybody was selling. You can profit in the same manner. Sell timber when you are besieged by buyers. Plant trees and buy land when other tree farmers are discouraged by the temporarily poor market. Every time the sun comes up, the market for tree farm products improves.

Indications are, however, that demand for wood will increase with the growing population. The U.S. Forest Service and some large manufacturers continually study future demands, and, although their estimates may differ, all predict increases. The combination of increased demand for wood and a shrinking area on which to produce it appears to foretell a market better than all previous ones. Now is the time to sow the seeds for the timber fortunes of tomorrow.

TREND TOWARD MANUFACTURING PLANTS WITHOUT LAND BASE

I predict an acceleration in the present trend toward manufacturing plants that depend on individual tree farmers to grow necessary raw material. I believe this is a beneficial trend.

First, it is efficient use of capital for the industrial concern. Its business is the manufacture and sale of useful products from wood, and concentration on the main effort increases its profit. Its income does not suffer the drag of carrying tree farms that produce a low return. Its capital requirements are smaller, since that required to grow raw material is supplied by individual investors. It can invest some of these savings in research on new products and processes, thereby increasing its profit and the market for tree farm products at the same time.

Second, smaller capital requirements mean that more mills will be built, assuming that markets develop as everyone predicts. Money is always hard to find, and it is easier to find small amounts than large amounts. Tree farms are particularly difficult to finance, since the business is new and they do not always produce regular annual incomes, which make financing easier. An adequate supply of raw material offered on the open market by local tree farmers eliminates one major problem of a board of directors planning construction of a manufacturing plant.

REAPPRAISAL OF MANY EXISTING FORESTRY PROGRAMS

I predict that many present forestry programs will be discontinued either because they were a waste of time in the first place or because they are no longer necessary.

Foresters have been persistently preoccupied with the problems of small landowners, whose tree farms are largely unproductive because of bad management and small amounts of capital. Educational programs were inaugurated to convince these people of the advantages of good forestry, and this approach was supposed to cure the problem. This was always an impossibility. You cannot make a capitalist by education alone, and you cannot make a good forest manager out of a man with six small children trying to scratch out a living on a 200-acre farm. The market will correct this situation, since more capable and experienced investors will soon acquire these lands by continuing to make a profit every year. Trying to educate people who are financially unable to use the knowledge is pouring money down the drain.

Many industries and some governments hired foresters to teach good forestry by performing various forest-management services either free or for nominal charges. Such programs have been phenomenally successful, as a review of Southern Pulpwood Conservation Association activities shows, but, once the lesson has been learned, tree farmers do not want unnecessary services. In fact, they are suspicious of free programs; everyone expects to pay for services if they are worthwhile. Tree farmers suspect they are paying for such assistance, and they are afraid because the price tag is not clearly visible. Industries in general have been quick to recognize this change in attitude and have reacted by placing increased emphasis on other efforts to make tree farming more profitable for individuals. Governments have not been equally quick, but increasing demands on tax revenues may force this realization on them.

Reappraisal of these programs may improve the outlook for tree farmers. Emphasis may be placed on assisting the investor who has sufficient capital to function as a tree farmer. Governments may decrease their expenses by eliminating needless programs or may improve the climate for tree farming by channeling funds formerly spent for education into programs of increased fire protection. Industries may improve their markets and the markets for tree farm products by directing similar funds into research to develop new products. Both may increase expenditures for research on the science of tree farming. Research of all kinds is impossible for most tree farmers, and this is an area where help is needed.

17

Case History of a
Successful Tree Farmer:
Charting a Course

In preceding chapters, I have discussed the important aspects of the business of tree farming. Now let me summarize by listing in order the steps you should take along your way to success; I shall do this by presenting a hypothetical case history of the operations of Robert Bryant.

Bryant, an independent oil operator, purchased minerals, royalty, and leases all over his state beginning in the 1930s and, when the price appeared to be cheap, bought the surface also. In this manner, he acquired in 1952 a tract of 2130 acres 120 miles southeast of his headquarters. At the time of the purchase, he gave only minor consideration to its operation as a tree farm. Rentals from mineral leases produced an adequate return on his investment; ad valorem taxes were low; most merchantable timber had recently been removed. By 1965, however, all of these conditions have changed, and sound money management makes it necessary to explore other possibilities. What steps does he take?

STEP 1: CHOOSE FORESTRY ADVISER

Unfamiliar with the services available to tree farmers, Bryant calls on the state forestry department, discovers that its services to private landowners are mainly educational, and receives a list of consulting foresters practicing in the state. Using criteria described in Chapter 3, he selects one and invites him to discuss his problems. After a brief explanation of what he owns and what he wants, he asks the forester to suggest a solution and to discuss the fee schedules and working arrangements of consulting foresters. The forester says that his firm offers a complete line of services to tree farmers (also described in Chapter 3), that his fees vary with the type of work, and that he prefers to undertake the work one step at a time, without a long-term forest-management contract. This gradual approach pleases Bryant; he knows little about tree farming and consulting foresters, but he realizes how important the relationship may be. After an hour of discussion, Bryant is satisfied with the forester's competence and commonsense and chooses him as a forestry adviser. This decision is probably Bryant's most important one; as we shall see, Bryant has an asset of large size and relies heavily on his forester's advice when making decisions of long-term consequence.

STEP 2: TAKE TREE FARM INVENTORY

As a start in what both hope will be a long and profitable association, the forester suggests that Bryant take a tree farm inventory. The tract may have no possibilities for tree farming (and this will end their association); at any rate, they cannot make sound plans until they know exactly what Bryant has. The forester recommends a 10% inventory, describes what data will be gathered, and quotes a fee on a per-acre basis. On being assured that this fee also covers whatever time is necessary for a planning conference after the report is submitted, Bryant orders him to take the inventory. Ten days later, the forester submits his report as follows:

HOWARD, STEPHENS AND DOUGLASS
Consulting Foresters COURT SQUARE SOUTH BUILDING
ANYTOWN, STATE

14 April 1965

Mr. Robert Bryant
P. O. Box 187
Anytown, State

Dear Mr. Bryant:

We have just completed a tree farm inventory of your property described as SW¼, W½ of NW¼, Section 26; Entire less NW¼ of NW¼, Section 27; S½ of NE¼, SE¼, Section 28; E½, Section 33; W½, NE¼, E½ of SE¼, Section 34; W½ of NW¼, W½ of NE¼ of NW¼, SW¼ of SW¼, Section 35, all in Township 5 North, Range 10 West, and NW¼ of NW¼ less West 10 acres, Section 2, Township 4 North, Range 10 West, containing 2130 acres. Our report is presented below and on the attached map (page 162).

Specifications and Sampling Procedure

We tallied as sawtimber all pine trees 9.0 in. and up DBH and all hardwood trees 11.0 in. and up DBH and as pulpwood all trees over 5.0 in. DBH but smaller than sawtimber. We divided the trees into 2 in. DBH classes and estimated volumes to the following top diameters inside bark: pine sawtimber 7 in., hardwood sawtimber 9 in., pulpwood 4 in. Pulpwood volumes include standing trees only; we made no estimate of pulpwood volume in tops of sawtimber trees. We tallied all pine species as pine, all gums as gum, all oaks as oak, and yellow poplar as poplar.

Sawtimber inventory is based on a 10% strip sample; we passed twice though each 40-acre block and, on each pass, tallied the sawtimber trees on a strip 66 ft. wide. Pulpwood inventory is based on a 2% strip sample; we tallied trees of this size on 1/10-acre plots spaced five chains apart on the cruise line. Number of pine trees in the 4 in. DBH class were tallied on the pulpwood plots. Growth data were obtained by taking an increment core from the tree nearest the center of each pulpwood plot.

Timber Inventory

Our estimate of the total number and volume of trees by DBH class, species, and product is as follows (see page 158):

Pulpwood volumes are given in unpeeled cords of 128 cubic feet each, and sawtimber volumes are given in thousands of board feet, Doyle scale. Four-in. trees are listed under pulpwood although they are below merchantable size. Hardwood trees occur on the tract only in scattered strips of pine-hardwood type (see map); therefore, we do not believe we took a sufficient sample of them to produce a reliable estimate. Because of their small volumes and low values, this is not a serious error.

Trees smaller than the 4 in. DBH class are present in considerable numbers under stands of larger trees; most of them became established at the same time as the larger trees and are small now because they are overcrowded or overtopped. They will not make a significant contribution to future growth. Your problem is either too many trees per acre or none at all. Average volume per acre is low because trees are small and because many acres are idle or used for other purposes. We divided the tract arbitrarily into seven blocks during inventory, and volumes for each block are given on sheets following the map. (See pages 163-169).

Scattered all over the tract are patches of old pine stumps suitable for distillate wood. We think their volume is between 500 and 1000 tons; they are concentrated enough to be merchantable.

Forest Types

The map shows the location of the two main forest types on your tract. One is pine and is almost a pure stand of longleaf pine. The other is pine-hardwood made up of 50% pine and 50% hardwod. In this type the pine species is loblolly, and the hardwood species groups are those shown under timber inventory.

Pulpwood

DBH	Pine Trees	Pine Vol.	Soft Hardwood	Hard Hardwood	Misc. Hardwood	Total Trees	Total Vol.
4	58,860					58,860	
6	33,950	1,256				33,950	1,256
8	27,750	2,553				27,750	2,553
		3,809					3,809

Sawtimber

DBH	Pine 1	Pine 2	Pine 3	Oak 1	Oak 2	Oak Vol.	Gum 1	Gum 2	Gum Vol.	Poplar 1	Poplar 2	Poplar Vol.	Misc.
10	5,710	7,350											
12	1,280	2,650	1,700	90	40	4.7	130	140	10.6	30	40	2.8	
14	300	710	450	80	50	8.3	50	110	11.6	20	50	5.1	
16	160	380	230	20	60	9.0		40	5.0		30	3.8	
18	40	180	210								10	1.8	
20	10	20	10										
Total			21,390			22.0			27.2			13.5	0.0

Total

DBH	Pine Trees	Pine Vol.	Oak Trees	Oak Vol.	Gum Trees	Gum Vol.	Poplar Trees	Poplar Vol.	Misc. Trees	Misc. Vol.	Total Trees	Total Vol.
10	13,060	378.8									13,060	378.8
12	5,630	322.5	130	4.7	270	10.6	70	2.8			6,100	340.6
14	1,460	144.1	130	8.3	160	11.6	70	5.1			1,820	169.1
16	770	109.7	80	9.0	40	5.0	30	3.8			920	127.5
18	430	95.2					10	1.8			440	97.0
20	40	10.2									40	10.2
	21,390	1,060.5	340	22.0	470	27.2	180	13.5	0	0.0	22,380	1,123.2

Growth Study
_____

In calculating timber growth on your property, we ignored the hardwood component; it contains only about 5% of the volume and less of the value. We estimate annual growth of pine to be 173,000 board feet of sawtimber and 302 cords of pulpwood, and we believe you can remove this volume each year without dipping into timber capital. This very high *percentage* rate comes about because many pulpwood trees are growing into sawtimber and because all present sawtimber trees are relatively small. It will gradually drop if you do not remove each year's growth.

Growth rates vary widely by DBH classes; 12-in. trees grow better than 8%, 20-in. trees less than 3.5%. There is also a great difference within DBH classes depending on condition of individual trees. Well-spaced, vigorous trees grow rapidly; suppressed trees grow hardly at all.

Soil Analysis
_____

We did not make a thorough study of the soils, but information available from the U.S. Soil Conservation Service indicates that the site index for your pine areas is 80 and that for the pine-hardwood areas is 90. This is a very general statement, but indicates that you have no serious problems here and will do for the time being. At present, your major concern is making the best possible use of existing stands, and measurements of growth in the field will provide all necessary guides. We recommend more thorough study of your soils when markets and financial resources make possible more intensive forest management.

Markets and Market Prices
_____

Markets for all forest products are excellent. Within 30 miles of the tract, there are 10 sawmills, 7 pulpwood dealers, and 3 pole buyers, and competition among them is spirited. We estimate stumpage values for pine pulpwood to be $6.00 per cord and for sawtimber per MBF to be as follows: pine $40.00, gum and oak $20.00, and poplar $35.00. Stumpage value of distillate wood is $2.00 per ton. There is an insignificant volume of hardwood pulpwood—so small that we did not include it in the inventory—which has a market value of $1.00 per unit and can be expected to double or triple within the next 10 years. Unfortunately, it can never contribute substantially to income.

Fire
_____

Except for pine-hardwood areas too wet to burn, the entire tract burned over this past winter, and there are signs that this is a common occurrence. Since almost every tree is a longleaf pine, there is little loss from death of merchantable trees. On the other hand, there is a definite growth loss each year because many trees are defoliated by fire, and constant burning prohibits establishment of young pines in many areas that are now idle. These idle acres, inherently productive, are scattered all over the tract in irregular strips and patches which show on aerial photographs but are much too complicated to show on the map. We believe they can be made productive quickly and cheaply by planting, but no planting should be done until the fire problem is solved. The fires appear to be of incendiary origin to provide early spring grazing for numerous cattle in the area, and although your county is under the state fire control program, several determined firebugs completely overpower all fire suppression organizations. This is a major problem; we estimate your timbered land is at least 30% idle because of fire.

Development Work

Planting is needed on about 600 acres, but should not be undertaken until you can control fire. No TSI work is needed now or in the near future; constant burning has eliminated all pine reproduction and almost all hardwood from pine areas. Thinning is desirable on 480 acres as shown on the map. We estimate that a thinning from them will produce about three cords of pulpwood per acre now.

Trespass

All merchantable timber has been removed within the last year from about 35 acres in the southwest corner of SW¼ of SE¼, Section 33. This appears to be trespass; we found no evidence of an established boundary line either south or west of cut area. Mr. Rice, mentioned below, knew nothing about cutting.

Adverse Possession and Other Use

The gravel pit in NE¼ of SW¼, Section 26, occupies five acres, is abandoned, and is being recaptured by pine seedlings.

The pasture north of Highway 42 occupies forty acres, is heavily grazed, but receives no cultivation. These conditions also apply to the 15-acre pasture on east side of NE¼ of SE¼, Section 34.

The cultivated fields in southeast portion of tract occupy 80 acres and have been prepared for planting this year. The two barns are in use. All four houses on the tract are occupied, and Julius Rice, General Delivery, Anytown, lives in the one in southeast corner of SE¼ of SE¼, Section 34. Mr. Rice told us that he leases the pastures, fields, and houses from you and subleases some facilities to others. He said that all fences shown are yours and necessary for his operations; if not, they may be evidence of adverse pssession.

Rights of Way

The map shows rights of way we found and measured on the ground. The largest is that of ABC Power Company it is 100 ft. wide, about 8000 ft. long, and contains 18.6 acres. The DEF Pipeline Company ROW is 25 ft. wide, about 3000 ft. long, and contains 1.7 acres; there is also a telephone or telegraph line down its center. There are three ROW's of the local electric power association, they are 20 ft. wide, a total of 16,830 ft. long, and contain 7.7 acres. The ROW for State Highway 42 is 60 ft. wide, about 10,560 ft. long, and contains 14.5 acres. The only other road ROW's that reduce timber-growing area are those in the extreme southeastern portion and a short stretch just south of the highway; they are 30 ft. wide, about 4500 ft. long, and contain 3.1 acres. The total area used for all ROW's is 45.6 acres; there may be additional areas affected if any power-line ROW agreement permits cutting of danger trees.

Boundary Lines

Some boundary lines are established by fences shown on map; these fences appear to be accepted by adjoining owners. In the northern portion, you have at least one mile of common boundary line with XYZ Paper Company; these two strips of line are painted white and maintained by the company. Although we investigated at several points, we found no evidence of established lines on the west side of the land in Sections 28 and 33 or the south side of the land in Sections 33 and 34; this situation may have led to the cutting, and we believe these must be surveyed.

Please call on us if we can give you any further information or assistance. After you have a chance to study these data, we will be happy to help you plan and execute those programs you select.

Respectfully submitted,

Howard, Stephens and Douglass
CALVIN R. DOUGLASS

CRD:bn
Attachments

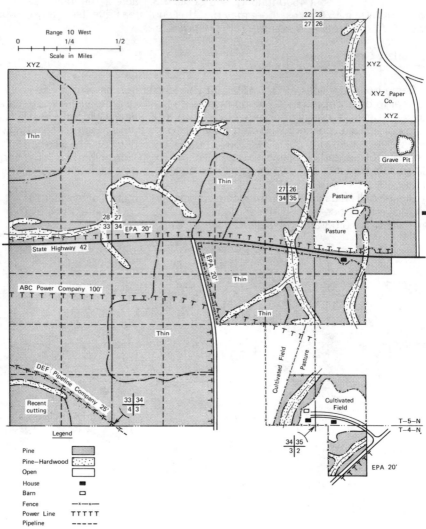

Figure 3

Block 1: W½ of NW¼, Section 26; NE¼, E½ of NW¼, Section 27

320 Acres

NUMBER OF MERCHANTABLE TREES BY DBH AND LOG LENGTHS

Pulpwood

	Pine		Soft Hardwood		Hard Hardwood		Total	
DBH	Trees	Vol.	Trees	Vol.	Trees	Vol.	Trees	Vol.
4	10,240						10,240	
6	5,600	207					5,600	207
8	4,250	391					4,250	391
		598						598

Sawtimber

	Pine						Oak				Gum				Poplar				Misc. Hardwood			
DBH	1	2	3	4	5	Vol.	1	2	3	Vol.	1	2	3	Vol.	1	2	3	Vol.	1	2	3	Vol.
10	750	830				44.9																
12	250	450	300			56.8					30	10		1.4								
14	60	130	80			26.4						20		1.6								
16	50	120	70			34.1						20		2.5								
18		40	70			26.2																
						188.4				0.0				5.5				0.0				0.0

	Pine		Oak		Gum		Poplar		Misc.		Total	
DBH	Trees	Vol.	Trees	Vol.	Trees	Vol.	Trees	Vol.	Trees	Vol.	Trees	Vol.
10	1,580	44.9									1,580	44.9
12	1,000	56.8			40	1.4					1,040	58.2
14	270	26.4			20	1.6					290	28.0
16	240	34.1			20	2.5					260	36.6
18	110	26.2									110	26.2
	3,200	188.4	0	0.0	80	5.5	0	0.0	0	0.0	3,280	193.9

Block 2: E½, Section 33
320 Acres
NUMBER OF MERCHANTABLE TREES BY DBH AND LOG LENGTHS

Pulpwood

DBH	Pine Trees	Pine Vol.	Soft Hardwood Trees	Soft Hardwood Vol.	Hard Hardwood Trees	Hard Hardwood Vol.	Total Trees	Total Vol.
4	10,020						10,020	
6	5,600	207					5,600	207
8	3,100	285					3,100	285
		492						492

Sawtimber

Number of logs

DBH	Pine 1	Pine 2	Pine 3	Oak 1	Oak 2	Oak 3	Gum 1	Gum 2	Gum 3	Poplar 1	Poplar 2	Poplar 3	Misc. Hardwood 1	Misc. Hardwood 2	Misc. Hardwood 3
10	500	570													
12	140	340	200	30				30	30						
14	50	130	80	20				20			30				
16	30	60	40		20			10							
18	10	60	60												

DBH	Pine Trees	Pine Vol.	Oak Trees	Oak Vol.	Gum Trees	Gum Vol.	Poplar Trees	Poplar Vol.	Misc. Trees	Misc. Vol.	Total Trees	Total Vol.
10	1,070	30.5									1,070	30.5
12	680	39.2	30	0.9	60	2.3					770	42.4
14	260	25.8	20	1.0	20	1.6	30	2.5			330	30.9
16	130	18.4	20	2.5	10	1.3					160	22.2
18	130	28.7									130	28.7
	2,270	142.6	70	4.4	90	5.2	30	2.5	0	0.0	2,460	154.7

Block 3: SE¼, E½ of SW¼, Section 27

240 Acres

NUMBER OF MERCHANTABLE TREES BY DBH AND LOG LENGTHS

Pulpwood

DBH	Pine Trees	Pine Vol.	Soft Hardwood Trees	Soft Hardwood Vol.	Hard Hardwood Trees	Hard Hardwood Vol.	Misc. Hardwood 1	2	3	Total Trees	Total 1	2	3
4	5,940									5,940			
6	3,300	122								3,300		122	
8	3,750	345								3,750			345
		467											467

Sawtimber

Pine

DBH	1	2	3	4	5
10	680	1,000			
12	160	330	200		
14	10	40	20		
16		30			
18		10	10		

Oak / Gum / Poplar / Hard Hardwood / Misc.

DBH	Oak 1	2	3	Oak Vol.	Gum 1	2	3	Gum Vol.	Poplar 1	2	3	Poplar Vol.	Hard Hardwood 1	Misc. Hardwood 1	2	3
10													20			
12	10			0.3	10			0.3		20		0.6	10			
14	10			0.5						10		0.5				
Total	20			0.8	10			0.3				1.1				

Totals by Species

DBH	Pine Trees	Pine Vol.	Oak Trees	Oak Vol.	Gum Trees	Gum Vol.	Poplar Trees	Poplar Vol.	Misc. Trees	Misc. Vol.	Total Trees	Total Vol.
10	1,680	49.6									1,680	49.6
12	690	39.3	10	0.3	10	0.3	20	0.6			730	40.5
14	70	7.0	10	0.5			10	0.5			90	8.0
16	30	4.2									30	4.2
18	20	4.6									20	4.6
Total	2,490	104.7	20	0.8	10	0.3	30	1.1	0	0.0	2,550	106.9

Block 4: W½, Section 34

320 Acres

NUMBER OF MERCHANTABLE TREES BY DBH AND LOG LENGTHS

Pulpwood

DBH	Pine Trees	Pine Vol.	Soft Hardwood Trees	Vol.	Hard Hardwood Trees	Vol.	Misc. Hardwood 1	2	3	Total Trees	Vol.
4	14,260									14,260	
6	8,050	298								8,050	298
8	6,600	607								6,600	607
		905									905

Sawtimber

DBH	Pine 1	2	3	4	5	Vol.	Oak Trees 1	2	3	Vol.	Gum Trees 1	2	3	Vol.	Poplar 1	2	3	Vol.
10	1,000	1,320	130															
12	100	210	10									20		0.9				
14	20	40	10									10		0.8				
16	20	20																
18	10																	
						0.0				0.0		30		1.7				0.0

DBH	Pine Trees	Vol.	Oak Trees	Vol.	Gum Trees	Vol.	Poplar Trees	Vol.	Misc. Trees	Vol.	Total Trees	Vol.
10	2,320	67.5									2,320	67.5
12	330	25.2			20	0.9					460	26.1
14	70	6.4			10	0.8					80	7.2
16	20	2.8									10	2.6
	2,860	104.5	0	0.0	30	1.7	0	0.0	0	0.0	2,890	106.2

Block 5: NE¼, E½ of SE¼, Section 34; SW¼ of SW¼, Section 35; NW¼ of NW¼ less West 10 Acres, Section 2

310 Acres

NUMBER OF MERCHANTABLE TREES BY DBH AND LOG LENGTHS

Pulpwood

DBH	Pine Trees	Pine Vol.	Soft Hardwood Trees	Soft Hardwood Vol.	Hard Hardwood Trees	Hard Hardwood Vol.	Misc. Hardwood 1	2	3	Total Trees	Total Vol.
4	4,100									4,100	
6	2,800	104								2,800	104
8	2,600	239								2,600	239
		343									343

Sawtimber

Pine

DBH	1	2	3	4	5	Vol.
10	350	1,080				45.9
12	130	270	200			34.8
14	50	140	100			29.3
16	40	70	60			24.4
18	10	30	40			17.6
						152.0

Soft Hardwood — Oak

DBH	1	2	3	Vol.
12	40	40		3.1
14	40	40		5.4
16	20	40		6.5
				15.0

Soft Hardwood — Gum

DBH	Trees	Vol.
12	140	5.6
14	110	7.5
16	10	1.3
	260	14.4

Hard Hardwood — Gum

DBH	1	2	3
12	60	80	
14	50	60	10

Hard Hardwood — Poplar

DBH	1	2	3	Vol.
12	10	30		1.7
14	10			0.5
16	10		10	1.3
18			10	1.8
				5.3

Total

DBH	Pine Trees	Pine Vol.	Oak Trees	Oak Vol.	Gum Trees	Gum Vol.	Poplar Trees	Poplar Vol.	Misc. Trees	Misc. Vol.	Total Trees	Total Vol.
10	1,430	45.9									1,430	45.9
12	600	34.8	80	3.1	140	5.6	40	1.7			860	45.2
14	290	29.3	80	5.4	110	7.5	10	0.5			490	42.7
16	170	24.4	60	6.5	10	1.3	10	1.3			250	33.5
18	80	17.6					10	1.8			90	19.4
	2,570	152.0	220	15.0	260	14.4	70	5.3	0	0.0	3,120	186.7

Block 6: SW¼, Section 26; W½ of NW¼, W½ of NW¼ of NE¼, Section 35

260 Acres

NUMBER OF MERCHANTABLE TREES BY DBH AND LOG LENGTHS

Pulpwood

DBH	Pine Trees	Pine Vol.	Soft Hardwood Trees	Soft Hardwood Vol.	Hard Hardwood Trees	Hard Hardwood Vol.	Misc. Hardwood 1	2	3	Total Trees	Total Vol.
4	4,200									4,200	
6	2,600	96								2,600	96
8	2,400	221								2,400	221
		317									317

Sawtimber

DBH	Pine 1	Pine 2	Pine 3	Pine 4	Pine 5	Oak 1	Oak 2	Oak 3	Gum 1	Gum 2	Gum 3	Poplar 1	Poplar 2	Poplar 3	Misc. Hardwood 1	2	3
10	950	1,030															
12	200	500	320									10					
14	60	140	100			10						10					
16	20	40	30									10					
18	10	30	20														
20	10	10															

DBH	Pine Trees	Pine Vol.	Oak Trees	Oak Vol.	Gum Trees	Gum Vol.	Poplar Trees	Poplar Vol.	Misc. Trees	Misc. Vol.	Total Trees	Total Vol.
10	1,980	56.1									1,980	56.1
12	1,020	59.3					10	0.5			1,030	59.8
14	300	29.9	10	0.5			10	0.8			320	31.2
16	90	12.9					10	1.3			100	14.2
18	60	12.4									60	12.4
20	20	6.1									20	6.1
	3,470	176.7	10	0.5	0	0.0	30	2.6	0	0.0	3,510	179.8

Block 7: SW¼ of NW¼, W½ of SW¼, Section 27; SE¼, S½ of NE¼, Section 28
360 Acres

NUMBER OF MERCHANTABLE TREES BY DBH AND LOG LENGTHS

Pulpwood

DBH	Pine Trees	Pine Vol.	Soft Hardwood Trees	Soft Hardwood Vol.	Hard Hardwood Trees	Hard Hardwood Vol.	Misc. Hardwood Trees	Misc. Hardwood Vol.	Total Trees	Total Vol.
4	10,100								10,100	
6	6,000	222							6,000	222
8	5,050	465							5,050	465
		687								687

Sawtimber

Distribution by number of logs:

DBH	Pine 1	Pine 2	Pine 3	Pine 4	Pine 5	Pine Vol.	Oak	Oak Vol.	Gum	Gum Vol.	Poplar	Poplar Vol.
10	1,480	1,520				84.3						
12	300	550	350			68.0	10	0.3				
14	50	90	60			19.4	10	0.8	10	0.8		
16	20	40	30			12.9			10	1.3		
18	10	10				3.2						
20	10	10				4.1						
	4,530					191.9	20	1.1	20	2.1		0.0

Summary by species:

DBH	Pine Trees	Pine Vol.	Oak Trees	Oak Vol.	Gum Trees	Gum Vol.	Poplar Trees	Poplar Vol.	Misc. Trees	Misc. Vol.	Total Trees	Total Vol.
10	3,000	84.3									3,000	84.3
12	1,200	68.0	10	0.3							1,210	68.3
14	200	19.4	10	0.8	10	0.8					220	21.0
16	90	12.9			10	1.3					100	14.2
18	20	3.2									20	3.2
20	20	4.1									20	4.1
	4,530	191.9	20	1.1	20	2.1	0	0.0	0	0.0	4,570	195.1

Bryant now has the information he needs to evaluate his asset and plan for the future. Of course, every property is different and so is every owner, but this gives you an illustration of what a tree farm inventory should contain. You probably have some data on your property, but they may be incomplete or out-of-date. You should resist the temptation to economize by using such data; even under optimum conditions, you must make many assumptions and plans about the future, and plans based on a shaky foundation lead to trouble. Although economy is fundamental in tree farming, false economy in inventory may prevent sound decisions and doom the enterprise to failure.

STEP 3: MAKE FINANCIAL FORECAST

Bryant and his forester now sit down with paper and pencil to see what returns can be had, now and later, from operation of the property as a tree farm, and to determine whether the money that can be realized from its sale could be better used in some other way.

First, they look at income. Annual growth is 173 MBF of sawtimber and 302 cords of pulpwood, which, at the values in the report, can be sold for $8730. The forester states that he will mark the trees for cutting, sell them, and supervise logging operations for a fee of 10% ($873), leaving a net of $7857. Rental from the lease to Mr. Rice is $160 per year, bringing total income to $8017.

Next, they look at expenses. Last year's taxes were $700, and, although these are likely to increase gradually, they use this figure. The tract has 10 miles of boundary line, but 1 mile is already maintained by the XYZ Paper Company. The forester estimates that it will cost $75 per mile ($675) to brush out and repaint these and that this work should be done every nine years; average cost per year is therefore $75. No other expenses seem necessary at present; Rice leases and lives on part of the tract, and the forester, in marking timber or inspecting cutting, will be on the remainder frequently. Therefore, total annual expense is $775, and net annual income is $7242.

The forester estimates that the tract, without minerals, can be sold for $80 per acre ($170,400). If the value is $170,400, an income of $7242 produces an annual return of 4.3%. Bryant thinks that land prices will rise steadily, and this consideration makes him content with a 4.3% return for the present.

There are two possibilities for increased income. Rice pays $160 per year for the use of four houses, 137 acres of cultivable land, and 240 acres of fenced woods pasture; this seems a low price and may be increased. Substantial increases in timber growth will be possible when elimination of uncontrolled fire permits intensive forest management; the forester states that the tree farm is capable of producing a net annual timber harvest of over $22,000. Therefore, Bryant has a fair return and good prospects. Without them, he would sell the tract at this point; with them, he proceeds to the next step.

STEP 4: CHOOSE TAX ADVISER

Bryant, who well understands the great effect of taxes, chose his tax adviser, a certified public accountant, many years ago and consults him on all business operations. To continue his study, he asks this CPA to analyze projected income from a tax standpoint.

The total purchase price in 1952 was $31,950.00 ($15.00 per acre), and, because of Bryant's primary interest in minerals, $5.00 per acre were allocated to minerals, $10.00 per acre to land, and no value to timber. Therefore, Bryant has no depletion basis for timber, and timber sales will be all capital gains, provided they are made in such a way that they qualify under federal tax laws. The tract is located in a state that does not allow special treatment of capital gains. Bryant's income is subject to federal taxes of 40% and state taxes of 4%. The CPA calculates the annual remainder after taxes.

Here is Bryant's income subject to federal taxes:

Income			
Timber sales	$8730.00		
Less commission	873.00		
	———		
Net sales—capital gain	7857.00		
Capital gain divided by 2		$3928.50	
Rental income		160.00	
		———	
Total income			$4088.50
Expenses			
Boundary-line maintenance		75.00	
Ad valorem taxes		700.00	
		———	
Total expenses			775.00
			———
Net income			$3313.50

Here is his income subject to state taxes:

Income
Timber sales	$8730.00		
Less commission	873.00		
Net sales		$7857.00	
Rental income		160.00	
Total income			$8017.00
Expenses			
Boundary-line maintenance		75.00	
Ad valorem taxes		700.00	
Total expenses			775.00
Net income			$7242.00

His total tax bill is, therefore,

Federal:	$3313.50 × 0.40 = $1325.40
State:	$7242.00 × 0.04 = 289.68
	$1615.08

And the remainder after all taxes each year is $8017.00 (net income) − $1615.08 (taxes) = $6401.92. This seems a small return on an asset that can be sold for $170,400.00, and the CPA continues his investigation.

Since the original cost of the land alone was $21,300.00, a sale at $170,400.00 will produce a capital gain of $149,100.00; subtraction of state taxes of 4% ($5964.00) and a maximum federal tax of 25% ($37,275.00) leaves an after-tax sum of $127,161.00. Bryant has two alternate investments for this sum. One is a tax-free municipal bond paying 4.5%, which will produce $5722.25 annually, less than the after-tax net income from the tree farm. Another is a 6% note, which will produce $7629.66 annually but will be subject to a 40% federal tax and a 4% state tax, leaving a remainder of only $4272.60.

An investment's value depends on how much money is produced after taxes; consequently, the tree farm has more merit for Bryant than appears at first glance. His return is better than that from another investment, and possibilities for much greater future profits are good. Now convinced that he has an attractive investment, he proceeds to the next step.

STEP 5: CHOOSE LAWYER

At the time of purchase, Bryant thought the price was cheap, and, since he knew the previous owner was a careful businessman, and since income from minerals was high, he asked his lawyer, an experienced associate for many years, to make only a hasty examination of the title, concentrating on minerals. Now, however, the value has appreciated to a large sum; tree farm operations appear likely to produce large incomes now and later; a tree farm inventory shows the existence and importance of many items formerly ignored. After a conference with Bryant and his forester and with a copy of the tree farm inventory in hand, the lawyer makes a thorough search of the title and, in his opinion, lists and discusses the exceptions.

First is the ROW conveyance to the ABC Power Company, dated July 5, 1928, by which the company acquired

. . . the right to construct, operate, and maintain electric transmission lines and all telegraph and telephone lines, towers, poles, and appliances necessary or convenient in connection therewith from time to time upon a strip of land 100' in width, as said strip is now located by the final location survey thereof heretofore made by said Company, over and across the lands of which it is hereinafter described as being a part, together with all rights and privileges necessary and convenient for the full enjoyment of use thereof for the purpose herein described, including *the right of ingress and egress to and from said strip* and the right to cut and keep clear all trees and undegrowth and other obstructions on said strip *and danger trees adjacent thereto where necessary.*

Said strip is a part of a tract of land described as follows: S½ of NE¼, Section 33, and S½ of NW¼, SW¼ of NE¼, NE¼ of SE¼, Section 34, all in Township 5 North, Range 10 West. Said strip is substantially described as follows: Fifty feet on each side of a line beginning at a point on the east line, said point being north 2,120 feet from the southeast corner of Section 34, thence N73°13'W 1730.0 feet to a point on south boundary of SW¼ of NE¼, Section 34, thence N73°13'W 1761.7 feet more or less to an angle point, thence N83°49'W 4,627.3 feet, more or less, to a point on west boundary line of S½ of NE¼, Section 33.

The lawyer points out the importance of the provisions I italicized above. The right of ingress and egress applies to all land described, not the ROW strip only, and may have some minor undesirable effect in the future; such all-inclusive access rights cannot be acquired by eminent-domain proceedings and should have been limited to the 100-ft. strip. The danger-tree-provision is necessary for the company, but it should have required the company to pay for the trees cut. In addition, the agreement contains no reversion clause. It is too late to correct these three errors. The agreement includes the exact location of the line, coinciding with that shown in the tree farm inventory; the lawyer states that the exact location should be part of every ROW agreement Bryant signs in the future.

The ROW agreement with the DEF Pipeline Company, dated June 20, 1930, is much simpler and has much more serious consequences. The former owner conveyed

. . . a right of way to lay, construct, maintain, and operate a pipe line or lines, consisting of one or more pipes, and appurtenances thereto, including telephone or telegraph lines in connection therewith, and the free right of ingress and egress to and from said right of way for the purpose of laying, constructing, maintaining, repairing, replacing, operating, or removing at will said pipe line and appurtenances hereto, and across the following described land: E½, Section 33, and W½, NE¼, Section 34, all in Township 5 North, Range 10 West.

Although the agreement also provides that the rights conveyed shall revert to the surface owner if they cease to be used, no location is stated, nor is there any mention of width. The company now uses only a 25-ft. strip containing less than two acres in the southwest portion of the tract; nevertheless, it has the right to operate more or less at will on all land described above, a total of 800 acres. This seriously limits Bryant's ability to make trades in the future and may hinder use of the land for residential or other development. Note here that the rights were sold in 1930. Bryant did not realize the existence or importance of this conveyance until 35 years later, and the ultimate importance may not be evident for another 20 years. The lawyer states that Bryant may have some remedy, and Bryant instructs him to proceed with it. (I shall omit further details of this because of the legal technicalities involved.)

The public records contain no highway or road ROW's, no easements for electric-power-association lines, and no lease to Rice. In response to an inquiry from the lawyer, the state highway department reports that its files contain no ROW conveyances across the Bryant tract. This being true, the lawyer states that the highway department, by adverse possession, has acquired at least an easement on the area involved. Therefore, Bryant cannot cut and use trees on the highway or road ROW's, but he may be entitled to the proceeds from their sale whenever the highway department sells them; the lawyer asks the forester to report any cutting on the ROW.

The electric power association provides copies of its easements, and the lawyer records them in the public records as additional evidence of Bryant's possession. Upon examining them, he finds no serious defects, and he notes that they do not allow cutting of danger trees; he asks the forester to report any cutting of this type, since it would be unauthorized.

The lawyer thinks that the arrangement with Rice is too informal, that Rice or another tenant may possibly bring about a case of adverse possession, and that a lease is desirable as evidence of possession. Furthermore, Rice's annual rental is low because he is supposed to supervise the entire tract; since he did not prevent or even know about the 35-acre trespass in Section 33, a more definite expression of this responsibility may be wise. Therefore, the lawyer prepares the following lease to be executed and recorded:

STATE OF_____
COUNTY OF_____

AGRICULTURAL LEASE

This lease entered into on this date by and between Robert Bryant, hereinafter referred to as Lessor, and Julius Rice, hereinafter referred to as Lessee. Lessor does hereby lease and let unto Lessee the following described lands, together with any and all improvements situated thereon:

South half of Southwest Quarter, Section 26; East half of Northwest Quarter east of road, Northeast Quarter south of highway, East half of Southeast Quarter, Section 34; West half of Northwest Quarter, West half of Northeast Quarter of Northwest Quarter, Southwest Quarter of Southwest Quarter, Section 35, all in Township 5 North, Range 10 West, and Northwest Quarter of Northwest Quarter less West 10 acres, Section 2, Township 4 North, Range 10 West.

The term of this lease shall be for a period of one year from the first day of January 1965 to the 31st day of December 1965 and shall automatically be extended each year for an additional term of one year unless either party shall give notice to the other party on or before the first day of December of his desire to terminate same.

The annual rental, in addition to the other considerations hereinafter set forth, shall be One Hundred Sixty Dollars ($160.00), payable on or before the 30th day of September.

It is mutually agreed between Lessor and Lessee:

1. Lessee will not use the leased lands for other than agricultural and residential purposes.

2. Any three of the four houses located on the leased premises may be sublet in the discretion of Lessee; however, Lessee must make his residence on the leased premises and all agricultural operations on the leased premises shall be carried on by Lessee.

3. Lessee shall not cut or deaden any trees of merchantable species, regardless of size. Lessor reserves unto himself, his successors and assigns the right of ingress and egress to and across the leased premises for the purpose of his tree farm and/or mineral operations on the leased premises and on lands adjoining the leased premises owned by Lessor.

4. Lessee agrees to maintain without expense to Lessor the buildings, fences,

and other improvements located on the premises in at least as good condition as when received by him.

5. The parties recognize that Lessor owns approximately 1800 acres of land adjoining the leased premises and as a part of the consideration for the execution of this lease by Lessor, Lessee agrees to make periodic inspections of these other lands as often and to the extent circumstances may indicate warranted, but in any event at least semiannually Lessee shall walk the perimeter thereof and report to Lessor any irregularities noted.

6. Lessee recognizes the title of Lessor to the leased premises and to the lands adjoining owned by Lessor and makes no claim to any of the said lands, adverse or otherwise, other than under the terms of this lease agreement.

7. Should Lessee fail to perform any of the covenants and agreements contained herein, including specifically but not limited to payment of the prescribed rental, Lessor may at his option cancel and terminate this lease and remove Lessee from the premises. In the event of termination of this lease for any breach by Lessee of any covenant or agreement contained herein, Lessee shall not be entitled to any rebate or adjustment for any unearned portion of rent accrued or paid to Lessor.

EXECUTED IN DUPLICATE ORIGINAL on this, the ＿＿＿＿ day of

＿＿＿, 1965.

＿＿＿＿＿＿＿＿＿＿＿＿＿＿

Lessor

＿＿＿＿＿＿＿＿＿＿＿＿＿＿

Lessee

(Acknowledgment forms omitted for brevity.)

There are no other title flaws, and, since those discussed can be either corrected or reduced in importance by adapting forest-management programs to them, Bryant proceeds to the next step, hand in hand with his advisers.

STEP 6: CHOOSE AND IMPLEMENT
FOREST-MANAGEMENT PLAN

Here is the meat in the coconut. As we said earlier, a management plan as part of tree farm inventory is premature; now is the time. A well-considered plan is essential to keep Bryant from running off in all directions at once, and he now has enough data and professional advisers to make a wise one.

He decides that a regular annual income is not desirable; he wants his tree farm to work as part of his investments, to provide income when other assets do not and opportunities for deductible expenses when other assets produce bountifully. Therefore, he chooses to do now those things that are obviously desirable and schedule future actions as they fit his needs and desires. His forester suggests four steps that are unquestionably desirable:

1. Sale of all old pine stumps suitable for distillate wood
2. Sale of all sawtimber growing at a compound rate of less than 6%, and of other sawtimber trees that are diseased, crooked, or otherwise defective (The forester states, and the inventory indicates, that this timber is located in the pine-hardwood areas and at their borders and that the sale will produce about $7500.)
3. Sale of all pulpwood that should be removed from the thinning areas, which will produce about $7700 and must be coordinated with sawtimber sales to prevent disputes between loggers
4. A survey of 2¾ miles of unestablished boundary line on the south and west sides of the tract, at an estimated cost of $400.00

The CPA states that Bryant can make either lump-sum or pay-as-you-go sales without jeopardizing capital gains treatment of his income. On the advice of the forester, he elects to sell sawtimber on a lump-sum basis and other products on a pay-as-you-go basis. The lawyer suggests that potential danger trees near the ABC Power Company ROW be cut regardless of growth rate, since they might be cut by maintenance crews without Bryant's knowledge.

Bryant chooses a well-known surveyor to establish the line and, after determining that his fee is the usual one, selects the forester to carry out timber and stump sales. The forester states that he will study the fire problem and suggest what might be done in the fall of 1965 to alleviate it. Bryant writes a memorandum confirming these decisions, and the management plan is complete. The following chapter discusses at length how it is executed and how such a scheme might be applied to your tree farm, even if it is a large industrial holding. Except in unusual cases, more elaborate plans are seldom necessary; the important thing is to consider all possible programs, with emphasis on the financial aspects of each. Doing so makes success likely. Tree farming does not require magic or genius—merely hard work and careful thought. As time passes, Bryant (and you) will be ready for the next two steps.

STEP 7: CHOOSE OTHER PROFESSIONAL ADVISERS

These will be needed only when the occasion arises, and Bryant's present advisers may be helpful in choosing others. Bryant is an expert on minerals, but you are likely to need a minerals adviser if you own minerals that are or become valuable. You may need a real estate broker if the locality becomes more heavily populated and the tree farm becomes more desirable for purposes other than tree farming. In certain cases, you will need a specialist in recreational developments.

Bryant has some cultivated land now, and there may be other portions that can be made suitable for farming. He will need advice from someone familiar with soils, their value for agriculture, and the cost of conversion. This advice may be valuable in renegotiating the lease with Rice, but he postpones action on this, since he is already committed for 1965.

STEP 8: REAPPRAISE THE TREE FARM WHEN NECESSARY, BUT AT LEAST ANNUALLY

Substantial new developments make reappraisal necessary whenever they occur, but an annual review of the tree farm investment, including progress of the past year and plans for the next year, is an excellent idea. Few assets produce properly when you buy them and forget them. Stockbrokers and bond dealers recommend constant review of your portfolio, and this is good advice for tree farmers. Trees grow; market prices change; new species become salable; population pressures on land vary; personal finances are altered. September is a good month for an annual review, since growth for the year is complete. At that time, income from other sources may be estimated closely, and yet there remains enough time in the calendar year to carry out actions to make tree farm operations fit more nearly into your entire investment program.

PROGRAM FOR WOULD-BE TREE FARMERS

Bryant's course (with suitable variations) can well be followed by all tree farm owners, but there are other factors to be considered if you are an investor anxious to become a tree farmer. You should follow the program outlined below.

First, you need to have a tree farm, several preferably, offered to you. Do not choose one that suits your fancy and then try to buy it. Enticing an owner to part with his property invariably means that you will pay or offer to pay him more than he thinks it is worth, and he is right much more often than you are. You should look for those who want to sell and are looking for you. An excellent means of locating them is a small advertisement in the classified section of a newspaper of wide circulation. You will often be surprised at the number of replies. You should also inform local real estate brokers and foresters of your desires; the first group will earn commissions by pleasing you, and the second group is often in touch with tree farmers anxious to sell.

Once you begin to receive offers, carry out Step 1, above. A good forestry adviser can help greatly in investigation of offers. Some tracts will be obviously worthless; some can be culled on cursory examination; a few may be worth inventory; and, finally, one will be worth buying. Your adviser can save you much time and expense at this stage.

Eventually, a property will appear so desirable that you will carry out Step 2. Do not be discouraged if data gathered in the inventory condemn the property as an investment. You will be out the cost of the inventory, but the knowledge you gain by a thorough investigation will be invaluable later and will improve your efficiency in appraising other offers. Under no circumstances should you purchase a property without a thorough investigation. All sellers try to create the impression that buyers are standing in line to get the property. Do not be stampeded; if you miss one good deal, another will be along shortly. Fast action may be essential at some times, but wise tortoises win more races than careless hares. During the time you are looking, your capital can be earning some return in other assets, whereas the losses caused by hasty purchases can be recovered only with difficulty. Perseverance will soon enable you to buy a tree farm that meets your requirements, and you can follow the entire program suggested above.

So far, I have not mentioned one great advantage of tree farming. You can do nothing—or at least nothing except pay taxes and prevent trespass and adverse possession, actions essential to maintain your ownership. Nature abhors a vacuum, and your tree farm will produce something no matter what you do. The advantage in this is twofold. First, a tree farm requires only a minimum of supervision by the investor; even when he does nothing, it will produce something. Second, a forest-management program started in prosperity can be postponed, reduced, or eliminated in adversity. Hard times come to all, and one hesitates to embark on a program of capital investment from which there is no escape. Tree farms, being dynamic, lend themselves well to variation in available investment funds; cultivation this year makes them more valuable, even if the program must be discontinued the following year. Moreover, values added by expenditures today may be realized in future times of need. Although maximum benefits come from continued cultivation of the crop, nature responds to the smallest amount of help. Now is the time to begin.

18

Case History of a Successful Tree Farmer: Executing the Plan

Bryant's plan, detailed enough for the present, is made; now let us follow its execution. No two tree farms or tree farm owners are alike, but perhaps his methods can be adapted to solve your problems whether your ownership is small or very large. We shall examine his actions in detail as they occur.

First, into an ordinary file folder Bryant puts all records he now has: deed, early reports and correspondence, tree farm inventory, title opinion, memorandum of forest-management plan, and so on. He will insist upon written reports of all future actions; all advisers urge that he make his record as he goes, and he will soon see the value of this advice. Next, he obtains another print of the inventory map, without shading this time, and affixes it to the inside cover of the file folder, facing the main body of material. He will use this to record as much data as possible and will hereafter be able to see a graphic picture of his operations merely by opening the file. We shall look at this map from time to time to note his progress.

SURVEYOR'S REPORT

The surveyor selected to establish 2¾ miles of southern and western boundary line submits a written report dated May 15, 1965, and accompanied by a sketch of the fieldwork. He states that he was able to prove the location of the southwest corner of E½, Section 28, and the southeast corner of Section 33 by the original government field notes and that he established pine-knot corners at these points and at the southwest corner of E½, Section

33. He finds no other items of interest except the cutting already reported in the inventory. Bryant files the report and sketch and marks on his status map the corners established and the year the work was done.

LUMP-SUM SALE OF SAWTIMBER

The first major activity is the lump-sum sale of sawtimber, and it is worth studying in detail from beginning to end. Bryant selects his forester to carry out this sale. The forester states that he will handle all parts of the sale for a 10% commission, that spirited competition in the area is likely to produce a suitable bid, but that he must charge Bryant a fee of $2 per MBF for his work if no acceptable bid is received. Bryant agrees, and the forester marks and estimates trees to be cut and sends the following sale announcement to 125 potential bidders located within a 75-mile radius:

HOWARD, STEPHENS AND DOUGLASS
Consulting Foresters COURT SQUARE SOUTH BUILDING
ANYTOWN, STATE
22 May 1965

Dear Sir:

Acting as agent for Robert Bryant, we solicit sealed bids on the timber marked for cutting on about 80 acres of his land in Sections 26, 27, 28, 33, 34, and 35, Township 5 North, Range 10 West, Any County, State. We have marked trees to be cut with yellow paint, and our estimate of their volume in board feet, Doyle scale, is as follows:

Pine	168,700
Gum	10,700
Oak	7,400
Poplar	12,500
Total	199,300

Since our reputation is at stake, we tallied the number of these trees with utmost care and estimated their volume in accordance with the most prevalent scaling practice in the area. We cannot guarantee them, however, because utilization practices vary too widely. Bidders should assume that they will pay any severance or documentary taxes due. The sale agreement will provide penalties to cover cutting or damage of unmarked trees, and a period of time until 30 June 1966 will be allowed to cut and remove the timber. Attached are a sheet showing the kind and size of trees marked (page 183) and a sample for the conveyance to be used (pages 182–184).

Only sealed bids will be accepted, and owner reserves the right to reject any or all bids. Bids should be mailed to Mr. Robert Bryant, c/o Howard, Stephens and

Douglass, P.O. Box 902, Anytown, State. Bids must be made on a lump-sum basis and must be received at the above address not later than 12:00 noon on 5 June 1965.

We will be at the traffic light at the intersection of State Highways 42 and 15 at 9:00 a.m. on 29 May 1965 to show interested parties over the area. The timber is located about four miles west of this point. Please feel free to call on us for further information or assistance.

Sincerely yours,

Howard, Stephens and Douglass
CALVIN R. DOUGLASS

CRD:bn
Attachments

STATE OF _____
COUNTY OF _____

TIMBER DEED

For and in consideration of the sum of One Hundred Dollars ($100.00), cash in hand paid, and other good and valuable considerations, the receipt and sufficiency of all of which are hereby acknowledged, ROBERT BRYANT, hereinafter called "SELLER," does hereby convey and warrant unto _____, hereinafter called "PURCHASER," all timber marked by Howard, Stephens and Douglass for cutting, as hereinafter indicated, on the following described lands:

West half of Northwest Quarter, Section twenty-six; South half, Section twenty-seven; South half of Southeast Quarter, Section twenty-eight; North half of Northeast Quarter, Section thirty-three; Northeast Quarter, East half of Southeast Quarter, Section thirty-four; and West half of Northwest Quarter, West half of Northeast Quarter of Northwest Quarter, Southwest Quarter of Southwest Quarter, Section thirty-five, all in Township Five North, Range Ten West, Any County, State.

All timber sold under this agreement has been marked with yellow paint spots below stump height and on the body of the trees. For any unmarked trees containing merchantable timber which are cut by Purchaser, his employees, contractors, or employees of contractors, Purchaser shall pay Seller at double the current price of stumpage for the class of material said trees contain.

No unnecessary damage shall be done to young growth or to trees left standing. Purchaser shall have the right of ingress and egress on, across and over the lands owned by Seller, except cultivated fields and open pasture lands, for the purpose of logging the timber conveyed herein; fences required to be cut for access shall be kept in place at all times and permanently repaired when such access is no longer needed. Purchaser may cut and use such small hardwood timber as may be necessary for bridging, roadbuilding, and logging.

Robert Bryant Tract
Sections 26, 27, 28, 33, 34, and 35, T5N, R10W, Any County
NUMBER OF MARKED TREES BY DBH, SPECIES, AND 16-FT LOG LENGTHS [a]

DBH	Pine					Oak			Gum			Poplar			Misc. Hardwood		
	1	2	3	4	5	1	2	3	1	2	3	1	2	3	1	2	3
10	28	31															
12	31	63	13			60			43	13		8	7				
14	32	137	133	13		42	3		19	27	1	5	20	7			
16	18	137	165	20		11	4		9	19	3		14	9			
18	3	81	102	10		8	1			6	3	1	8	9			
20	2	28	28			2	1		1				1	3			
22		10	6				1							2			
24	1	3	2											1			

DBH	Pine		Oak		Gum		Poplar		Misc.		Total	
	Trees	Vol.b	Trees	Vol.b	Trees	Vol.b	Trees	Vol.b	Trees	Vol.b	Trees	Vol.b
10	59	1.7									59	1.7
12	107	5.7	60	1.9	56	1.9	15	0.6			238	10.1
14	315	33.8	45	2.4	47	3.3	32	2.6			439	42.1
16	340	55.5	15	1.3	31	3.5	23	3.2			409	63.5
18	196	46.1	9	1.0	9	1.8	18	3.6			232	52.5
20	58	17.2	3	0.5	1	0.2	4	1.2			65	18.9
22	16	6.0	1	0.3			2	0.8			20	7.3
24	6	2.7					1	0.5			7	3.2
	1097	168.7	133	7.4	144	10.7	95	12.5	0	0.0	1469	199.3

[a] Volume is shown in thousands of board feet, Doyle Scale.

[b] Explanation: figures under "DBH" show size of tree; line of figures directly opposite "DBH" shows merchantable height of trees in number of 16-ft logs. Figures in main body of table give number of marked trees of each size, height, and species. Thus the numeral 133 in row 3 opposite 14 indicates that the marked timber includes 133 pine trees that are 14 in. DBH and contain three 16-ft logs.

Unless extension of time is granted in writing by Seller, the timber sold under this agreement shall be cut and removed from the above-described lands by June 30, 1966; title to any timber sold under this agreement and remaining on the lands described above after such deadline or any extension thereof shall revert to Seller.

Purchaser agrees and warrants that it will at all times indemnify and save harmless Seller against any and all claims, demands, actions or causes of action, for injury or death of any person or persons, or damage to the property of any third person or persons, which may be due in any manner to operations of Purchaser upon these lands.

WITNESS MY SIGNATURE on this, the _____ day of June 1965.

_____

Robert Bryant

(Acknowledgment forms, omitted here for brevity, are necessary to complete conveyance.)

On June 5, Bryant receives five bids ranging from $6675.75 to $8652.00 and accepts the highest. A sixth bid arrives two hours late and is returned unopened. Bryant's lawyer prepares the conveyance that afternoon; it is executed by both parties; the sale is closed before the end of the day. Bryant makes sure the conveyance is recorded and keeps a copy in the file; both actions add to his record of possession. He follows a similar procedure with all conveyances, leases, and contracts.

Bryant now plots this action on his status map; although the sale is both harvest and thinning, it is sufficient to show a partial cutting in 1965. In addition, while marking timber near the northeast corner of the tract, the forester discovered an additional mile of common boundary line with the XYZ Paper Company. It begins at the northeast corner of NW¼ of NW¼, Section 26, and runs due west one mile; except for a concrete monument at the section corner, there is a pine knot at each quarter-mile. Bryant plots this information also.

STUMP SALE

The sale of stumps for distillate wood is a simple matter. There are only three buyers, and a letter from the forester asking for bids per ton as harvested produces a high bid of $2.08. The lawyer prepares the conveyance. Because of the small dollar volume involved, no deposit is required to ensure completion; the sale proceeds simultaneously with that of sawtimber, because the two crews are not likely to interfere with each other. Since the stump sale is small and will never be repeated, Bryant does not plot it on the status map.

PULPWOOD SALE

The pulpwood sale is postponed until after June 30, 1966, to allow saw-timber loggers to vacate the area, thereby eliminating possible disputes about cutting unmarked trees. Since this delay is necessary, the advisability of the sale is reconsidered and approved again in September 1965, by Bryant, the CPA, and the forester. Therefore, the forester marks those trees to be cut and sends the following sale announcement to 14 potential pulpwood buyers in the area:

HOWARD, STEPHENS AND DOUGLASS
Consulting Foresters COURT SQUARE SOUTH BUILDING
ANYTOWN, STATE
10 July 1966

Dear Sir:

Acting as agent for Robert Bryant, we solicit sealed bids on the timber marked for cutting on his property in Sections 26, 27, 28, 33, 34, and 35, Township 5 Noth, Range 10 West. Trees to be cut are marked with blue paint; the area of marked timber contains about 480 acres; we estimate the volume to be 1350 unpeeled standard cords.

Bidders should assume they will pay any severance or documentary taxes due. At time of closing, owner will require a deposit of $500.00 to insure that all marked trees are cut and removed. Sale agreement will provide penalties to cover cutting or damage of unmarked trees, and a period of time until 30 June 1967 will be allowed to cut and remove the timber. Attached is a sample of the conveyance to be used [below].

Only sealed bids will be accepted, and owner reserves the right to reject any or all bids. The timber is suitable for pine pulpwood only and will be paid for weekly as harvested. Bids must be made on a per cord basis. They should be mailed to Mr. Robert Bryant, c/o Howard, Stephens and Douglass, P.O. Box 902, Anytown, State, and must be received at this address not later than 12:00 noon on 24 July 1966.

We will be at the traffic light at the intersection of State Highways 42 and 15 at 9:00 a.m. on 17 July 1966 to show interested parties over the area. The timber is located about four miles west of this point. Please feel free to call on us for further information or assistance.

Sincerely yours,

Howard, Stephens and Douglass
CALVIN R. DOUGLASS

CRD:bn
Attachments

STATE OF _____
COUNTY OF _____

PULPWOOD DEED

For and in consideration of the sum of One Hundred Dollars ($100.00), cash in hand paid, and other good and valuable considerations, the receipt of all of which is hereby acknowledged, ROBERT BRYANT, hereinafter called "SELLER," does hereby convey and warrant unto _____, hereinafter called "PURCHASER," all timber marked by Howard, Stephens and Douglass, for cutting as hereinafter indicated, on the following described lands:

South half, Section twenty-seven; South half of Northeast Quarter, Southeast Quarter, Section twenty-eight; East half of East half, Section thirty-three; West half, Northeast Quarter, Section thirty-four, all in Township Five North, Range Ten West, Any County, State.

All timber sold under this agreement has been marked with blue paint spots below stump height and on the body of the trees. For any unmarked trees containing merchantable material which are cut by Purchaser, his employees, contractors, or employees of contractors, Purchaser shall pay Seller at double the current price of stumpage for the class of material said trees contained.

No unnecessary damage shall be done to young growth or to trees left standing. Purchaser shall have the right of ingress and egress on, across and over the lands owned by Seller, except cultivated fields and open pasture lands, for the purpose of cutting and removing the timber conveyed herein; fences required to be cut for access shall be kept in place at all times and permanently repaired when such access is no longer needed. Purchaser may cut and use such small hardwood timber as may be necessary for bridging, roadbuilding and logging.

Unless an extension of time is granted in writing by Seller, the timber sold under this agreement shall be cut and removed from the above-described lands by June 30, 1967; title to any timber sold under this agreement and remaining on the lands described above after such deadline or any extension thereof shall revert to Seller.

Seller acknowledges receipt from Purchaser of the sum of Five Hundred Dollars ($500.00), to be returned to Purchaser when all marked trees have been cut and removed, or forfeited to Seller as liquidated damages should Purchaser fail to cut and remove all marked trees within the allotted time.

At the end of each week Purchaser shall pay to Seller for all material removed during that week at the rate of _____ per standard cord. No material shall be cut or removed during any time that such payments are in arrears without the written consent of Seller.

Purchaser agrees and warrants that it will at all times indemnify and save harmless Seller against any and all claims, demands, actions, or causes of action, for injury or death of any person or persons, or damage to the property of any third person or persons, which may be due in any manner to operations of Purchaser upon these lands.

WITNESS MY SIGNATURE on this, the _____ day of July 1966.

_____

Robert Bryant

(Acknowledgment forms, omitted here for brevity, are necessary to complete conveyance.)

Of the three bids received, Bryant accepts the high one, of $7.00 per cord, submitted by the Stephens Timber Company. The lawyer prepares the conveyance; Stephens puts up the deposit; the sale is closed the following day; and cutting begins immediately.

Because of good weather and steady orders for pulpwood, Stephens finishes cutting by the end of November; Bryant returns the deposit and enters the thinning on his status map. He also notes that Stephens owns SE¼ of NW¼, Section 28, and E½ of SW¼, Section 33.

At the annual review conference in September 1966, the forester states that there are no pressing problems in forestry or related matters, and Bryant and the CPA decide that no revenue other than that being produced by the pulpwood sale is necessary or desirable. Therefore, no action is scheduled until the meeting next year. Tree farms and people, however, are living things, and unforeseen problems arise. Two come up almost simultaneously, and both emphasize the importance of good records.

BOUNDARY-LINE DISPUTES

In May 1967 Rice, the tenant, and John Smith, owner of W½ of SE¼, Section 34, disagree about the location of the line between Smith's land and the cultivated field northwest of Rice's house—Smith contending that the proper location is 80 ft. east of the existing fence. Independently but simultaneously, H. E. Jones, owner of NW¼ of SW¼, Section 35, states that his western line is 10 ft. west of the existing fence, that 12 trees included in the 1965 sawtimber sale actually belonged to him, and that he wants to be paid for them.

Bryant searches his files and finds nothing bearing on either issue. His forester reports that both fences appear to be 10 or 15 years old, but he can suggest no way to prove this conclusively. His lawyer recommends that the disputes with Smith and Jones be settled by agreement and that all future acts of possession be carried out right up to the lines. Finally Bryant, Smith, and Jones select a surveyor and agree to accept the lines he establishes and to share his cost proportionately. The survey is made in the presence of Jones, Smith, and Bryant's forester and proves that Smith is wrong and Jones is right. Bryant pays Jones for the trees cut through error and puts all correspondence with Jones and Smith and the surveyor's report and plat in the file. The surveyor paints the lines established, and Bryant promptly enters as much of this information as possible on the status map. Better land management at earlier times could have prevented this needless expense, and Bryant resolves to practice it in the future.

TSI WORK

In September 1967 Bryant is near the end of a year when his income from other sources is much higher than normal; he wants no additional income and desires to invest a small amount of money in the tree farm if the expense is deductible and if the value added by the expenditure is sufficient to recommend the investment to a prudent man. His forester states that a fortunate absence of fire during the winters of 1965 and 1966 has allowed an excellent stand of loblolly pine reproduction to become established on about 80 acres north of the stream through N½ of NE¼, Section 33. These young seedlings are now three years old but are completely overtopped by a stand of worthless hardwood trees and unlikely to survive under this cover. He estimates the cost of TSI work to remove the hardwoods at $10 per acre ($800) and also estimates that, five years after removal, the accelerated growth of young pines will increase the value of the area treated by $20 per acre. The treatment area is protected from fire on the south and east sides by the stream, and the other two sides can be protected cheaply if necessary. The CPA states that such an expenditure is deductible, and Bryant orders the work done provided it is completed and paid for in 1967. With the help of the forester, he accomplishes this objective through a local contractor associated with the ASCS program, at a cost of $850. He promptly enters this action on the status map.

FOREST-MANAGEMENT RECORDS

Now let us pause and study Bryant's forest-management records through 1967. One of his first steps was to secure another print of the tree farm inventory map; the print was made from the original tracing and showed everything except the shaded forest-type symbols. After affixing the print to one cover of the file, Bryant devised another set of symbols, for forest-management activities—including one each for planting, first thinning, second thinning, and TSI work—and added this new section to the legend at the bottom of the page. Using these symbols in conjunction with dates, he plotted on the map in as much detail as possible every action taken and every item of information obtained. On page 190 is his status map at the end of 1967.

No one device is sufficient to show all needed information, but this map plus the file opposite it make a simple and highly effective tool. For example, if Bryant needs to know anything about the survey of his western line, a glance at the map shows that it was last done in 1965, and a quick turn in the file to the correspondence of that period reveals the action in detail.

Why is this necessary? First, Bryant is a busy man and cannot remember the myriad details of all his operations. His checkbook shows his bank account; his status map shows his tree farm; and both records permit a rapid review. His tree farm is a sizable asset, and he governs its disposition by the decisions he makes each September. Therefore, he needs something to show everything clearly and quickly, so that his decisions will be sound and cover all important aspects. Second, although he seldom inspects the property on the ground, he wants a ready reference to carry with him, so that his inspection trips will be efficient and thorough. The longer he operates the tree farm, the more actions he takes and the harder it is to grasp the whole picture by thumbing through a file. Third, he realizes that the actual fieldwork is done by others, that his experience is only secondhand. Since his relations with his advisers may be ended by death or disagreement, he wants a complete record to turn over to other agents. Gathering these data again would be expensive and nearly impossible. All these considerations make the status map of great importance and value to him, and you should follow his example on your tree farm, regardless of its size. Now let us continue our study of his actions.

TAXES AND MR. RICE

For several years after 1967, Bryant enjoys prosperous times and elects to do no cutting, allowing timber growth to accumulate for future use. Ad valorem taxes, originally nominal in amount, begin to rise sharply, however, and by 1972 reach 6.5% of assessed value annually. Bryant directs his attention to this serious matter and, for the first time, studies his assessment in detail.

He finds that uncultivable land is assessed at $15.00 per acre and cultivable land at $25.00 per acre and that he is assessed with 138 acres of cultivable land. He does not want to dispose of the cultivable land, but he can plant trees on it and reduce his assessment by changing the classification. He also notes that four houses and two barns have a total assessed value of $1800.00,

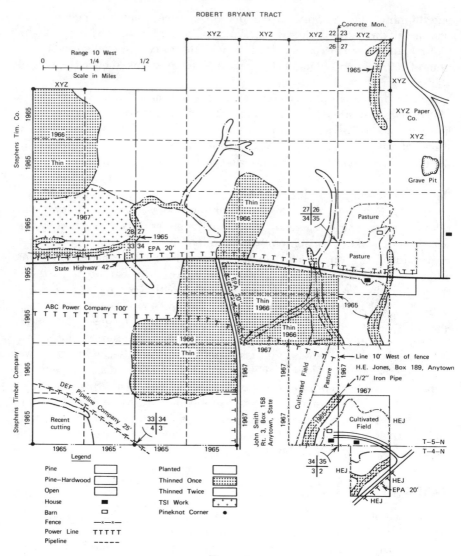

Figure 4

about $300.00 each. These installations are those leased to Rice, and Bryant calculates their cost to him in annual taxes, as follows:

Houses and barns	$1800.00 × 0.065 = $117.00
Excess assessed value of cultivable land	138 × $10 = $1380.00 × 0.065 = 89.70
	$206.70

Since Rice pays rent of only $160.00, the facilities he uses cause an annual loss to Bryant of $46.70. Local inquiry reveals that the buildings can be sold for $1800.00; when this amount, less state and federal taxes, is invested in a savings account, it will produce $84.00 annually. Bryant finds that he can sell the buildings, invest the net proceeds, convert the fields to timberland, obtain a reduction in ad valorem taxes, and end the year with $130.70 more cash than he does now with the Rice lease. In addition, he will be able to grow trees on about 138 acres now used by Rice. Further local inquiry reveals that the going rent for cultivated land is $10.00 per acre per year, and the forester reports that Rice cultivates 70 acres and grazes the remainder.

Bryant discusses all of this with Rice. Rice reports that the two houses on the highway and the barn north of the highway are almost unusable and that he can make little use of the pastures north of the highway. After much trading, they agree that Rice will relinquish these three buildings and the northern pastures, pay $700.00 per year for 70 acres of cultivated fields and $200.00 per year for use of the southern two houses and barn and the right to graze all land south of Highway 42 and east of the road through the middle of Section 34. Revision of the present lease confirms these changes. Bryant sells the relinquished houses and barn and applies for a reduction in assessment. He instructs the forester to submit a proposal to plant pine seedlings in the northern pastures, and he notes the sale of the houses and barn on the status map.

This demonstrates the value of constant attention to details, a prerequisite for successful tree farming. Bryant increases his annual cash income from leases by $740.00; he decreases his annual taxes by $84.50, or ($900.00 × 0.065 = $58.50) + (40 acres × $10.00 = $400.00 × 0.065 = $26.00); he has the net proceeds from the sale of houses and barn for other investments; he has 40 acres of old pasture for planting in pine seedlings. Similar action may be possible for you.

SECOND PULPWOOD THINNING

At the review conference in September 1972, the forester reports that the area thinned for pulpwood in 1966 now needs another thinning of similar kind and extent. Bryant and his CPA study his income for the year and determine that additional revenue from timber sales is desirable, and Bryant orders the thinning executed, provided all income is received in 1972. Since less than four months of the year remain, the forester suggests a lump-sum sale to ensure that income is received in time, but the CPA demurs, saying that this method may jeopardize capital gains tax treatment of timber income. This was not a danger when tree farm operations began in 1965, but the situation has gradually changed, and Bryant may now be classified as one who holds timber for sale to customers in the ordinary course of business. Therefore, the CPA strongly urges a pay-as-you-go sale, and the forester, intimately familiar with market conditions because of his daily operations, states that this is possible and carries out the order. He follows the same procedure used in the first thinning, and Bryant enters the action on his status map.

At the same conference, they discuss planting pine seedlings in the pastures relinquished by Rice. Money from the pulpwood thinning will be available, and they desire to put all land into productive use. The forester reports, however, that there is still no effective control of forest fires and that the seedlings are likely to be burned up soon after planting. They postpone action on this.

GRAZING LEASE

In 1973, a new state law requires cattlemen to prevent their cattle from wandering onto state highways, and this law, plus the steady development of the locality and the consequent shrinkage of available grazing area, offers Bryant an excellent opportunity. Local cattlemen request permission to erect fences on the north side of Highway 42 and on the south side west of the road in Section 34. The forester, convinced from his experience that the annual fires are set by cattlemen, is enthusiastic about the possibilities offered by this request. He suggests that everything not leased to Rice be leased to the cattlemen for grazing, that the lease require them to prevent or control all forest fires on the leased area (specifying that failure to do so would cancel the lease and require removal of the fences), that sufficient gates to allow access for timber operations be constructed, and that Bryant request an additional annual consideration of $450, about 25¢ per acre on the leased area. He also states that the cattle population is neither large enough to cause browsing damage to tree seedlings nor likely to become so, and that the grazing animals will reduce accumulation of fuel on the ground. After negotiation, the cattlemen agree to the fire-control provision and an annual rental of $250 in advance and select Edward Clark, one of their group, to receive the lease. After having his lawyer prepare it, Bryant executes and delivers the following lease:

STATE OF _____
COUNTY OF _____

GRAZING LEASE

This lease entered into on this date by and between Robert Bryant, hereinafter referred to as Lessor, and Edward Clark, hereinafter referred to as Lessee. Lessor does hereby lease and let unto Lessee the premises described as:

West half of Northwest Quarter, Southwest Quarter, Section 26; Entire less Northwest Quarter of Northwest Quarter, Section 27; South half of Northeast Quarter, Southeast Quarter, Section 28; East half, Section 33; West half less that part east of gravel road, Northeast Quarter north of Highway 42, Section 34; All north of Highway 42 in Northwest Quarter of Northwest Quarter and West half of Northeast Quarter of Northwest Quarter, Section 35, all in Township 5 North, Range 10 West.

The term of this lease shall be one year from January 1, 1973, to December 31, 1973. The annual rental shall be Two Hundred and Fifty Dollars ($250.00) in addition to the other considerations hereinafter set forth and shall be payable on or before the first day of the term. The lease shall automatically be extended for additional terms of one year each unless on or before the first day of December either party shall give notice to the other of his desire to cancel the same.

It is mutually agreed between Lessor and Lessee:

1. Lessee will not use said lands for any purpose other than livestock pasture. Lessee shall not cut, deaden, or remove any trees of merchantable species regardless of size.

2. Lessee may erect such fences as he may desire, but any fences erected shall be on and along the boundary line of the leased premises or rights of way of public roads. Fences erected along public roads shall contain at least two gates per mile at points to be designated by Lessor, and such gates shall be of a minimum width of ten feet.

3. Lessor reserves unto himsef ingress and egress on and across the leased premises for the purpose of his tree farm and/or mineral operations; in addition Lessor reserves unto himself the right to use and occupy the leased premises for any purpose or purposes not in conflict or inconsistent with the purposes of this lease.

4. Lessee will not himsef nor will he allow any other person to burn any grass, weeds, or other substance on said lands without prior consent of Lessor. It shall be the duty and obligation of Lessee to prevent and/or control all fires on the leased premises regardless of cause or source. In addition to all other rights and remedies of Lessor under this lease, if during any one year more than three hundred (300) acres are burned without prior consent of Lessor, Lessor may at his option cancel and terminate this lease and in such event it shall be the duty and obligation of Lessee to take down and remove all fences placed on the leased premises by Lessee.

5. Should Lessee fail to perform any of the covenants and agreements herein contained, including specifically but not limited to the payment of rent, Lessor may at his option cancel and terminate this lease and remove Lessee from the

premises. In the event of the termination of this lease for any breach by Lessee of any covenant or agreement herein contained, Lessee shall not be entitled to any rebate or adjustment of any rental paid or accrued.

6. Lessee may at his option sublet or assign all or any part of the privileges acquired by him hereunder, but such assignment shall not operate to release or relieve Lessee of any of his obligations hereunder. It is further agreed that no more than two hundred (200) head of livestock shall be grazed on the leased premises at any one time.

EXECUTED IN DUPLICATE ORIGINAL on this, the _____ day of _____, 1973.

_____

Lessor

_____

Lessee

(Acknowledgment forms omitted for brevity.)

Bryant plots the new fences on his status map. He hopes that such multiple use of the land will provide extra income and solve a serious forest-management problem.

OIL LEASE

In February 1975 the Gilchrist Oil Company, in an effort to complete a drilling block, requests an oil lease on the 30 mineral acres owned by Bryant under the land in Section 2. Gilchrist offers a bonus of $750 ($25 per acre) and annual rentals of $30 ($1 per acre) for a lease with a primary term of 10 years and tenders a 10-day draft for the bonus. Bryant, an expert on minerals, realizes that the only sure consideration is the bonus, since Gilchrist can cancel the lease at its pleasure, and he succeeds in getting Gilchrist to reduce the term to five years. They reach tentative agreement on all points, but Bryant says that he must confer with his lawyer before signing, since some provisions may be objectionable.

Gilchrist submits the lease on the printed form commonly known as the "Producers 88," and Bryant and his lawyer read it carefully. (I shall not reproduce this form because of its great length. It contains many technical details of mineral operations and should be discussed with your minerals adviser. You can get a copy from any oil operator and many office-supply companies in oil country.) The lawyer points out that it contains the "Mother Hubbard" sentence (see page 46 for discussion) and recommends that this be removed entirely. He also points out that it provides that "Lessee shall be responsible for all damages caused by Lessee's operations hereunder other than damages necessarily caused by the exercise of the rights herein granted." He recommends that Bryant put a period after "hereunder" and strike out all following words. Bryant is willing to accept the draft, but the lawyer recommends that the lease and draft remain attached until the draft is paid.

Gilchrist agrees to these changes, and Bryant executes the lease and deposits it and the draft at his bank for collection. To add to his record of possession, he requires Gilchrist to record the lease. No additions to the status map are necessary.

MNO PIPELINE COMPANY

In mid-1975, the MNO Pipeline Company approaches Bryant to acquire an ROW for a new pipeline through the northern portions of Sections 26 and 27. The buyer wants a 50 ft. ROW through NW¼ of NW¼, Section 26, and NE¼ of NW¼, N½ of NE¼, Section 27, and offers to pay $5.00 per rod for the ROW itself, plus whatever additional sums are necessary to compensate Bryant for timber on the ROW. The buyer states that he does not know the exact location of the ROW, because the location survey is not complete. Bryant delays decision and asks the buyer to return when the survey is complete.

Bryant uses the delay to prepare for the coming trading session. When a call to his lawyer reveals that MNO has the power of eminent domain and can take the ROW from him, he resolves to accomplish his objectives by trading and without a lawsuit, if possible. He calculates that $5.00 per rod for a 50 ft. ROW is equal to $264.55 per acre. His forester reports that, although the timber volume on the ROW cannot be accurately determined until the location is definite, the average value of merchantable timber on this part of the tract is $75.00 per acre. The forester also states that salvage of such a small volume is not economical. Since no minerals are to be conveyed, Bryant's depletion basis on what will be conveyed is only $10.00 per acre, so almost the entire proceeds will be subject to a state tax of 4% and a maximum federal tax of 25%. (Your tax adviser can suggest other, and perhaps more satisfactory, ways to handle depletion in ROW sales.) Furthermore, his annual ad valorem tax is $0.98 per acre, and he will be left with this burden on land he cannot use. Considering all these things, he decides to calculate his position if he settles with MNO for a total consideration of $10.00 per rod, or $529.10 per acre.

He has a capital gain of $529.10 − $10.00 = $519.10, which will be subject to maximum combined taxes of 29%, or $150.54, leaving a remainder of

$378.56. An investment of $17.50 at 6% will yield $1.05 annually, about enough to pay ad valorem taxes, and deduction of this leaves $361.06. The ROW contains perhaps $75.00 worth of merchantable timber and some smaller trees of undetermined value, but, after considering this, he thinks the compensation is adequate. Although he has spent some time in calculation and negotiation and some money on lawyers and foresters, he decides that he will sell for $10.00 per rod, or more if he can get it. When the buyer returns with the exact location, he and Bryant tentatively agree to trade at $10.00 per rod. The buyer asks that this be divided into $4.00 per rod for the ROW and $6.00 per rod for damages; Bryant agrees and asks him to prepare the necessary papers for review by his lawyer.

For the ROW conveyance, the buyer fills in the blanks on the usual company form, and, after checking the legal description and the amount of the consideration, Bryant hands it to his lawyer, who studies it carefully, finds that it contains many undesirable conveyances and agreements, and underscores all those words that should be removed. Here is the conveyance as he returns it to Bryant:

> For and in consideration of the sum of $1301.58, the receipt of which is hereby acknowledged, Robert Bryant, hereinafter referred to as Grantors (whether one or more), does hereby grant, bargain, sell, and convey unto MNO PIPELINE COMPANY, a Delaware corporation, its successors and assigns, hereinafter referred to as Grantee, a right of way and easement for the purpose of constructing, maintaining, inspecting, operating, protecting, repairing, replacing, changing the size of or removing a pipeline or pipelines, for the transportation of oil, gas, and the products or derivatives thereof, upon and along a route to be selected by Grantee across, over, through and under the following described lands, of which Grantors warrant they are the owners in fee simple, situated in Any County, State, to wit:
>
>> A strip of land twenty-five feet on each side of a line described as beginning at a point 330 feet north of the southwest corner of NE¼ of NW¼, Section 27, thence North 80° East 5361 feet to the east line of NW¼ of NW¼, Section 26, all in Township 5 North, Range 10 West,
>
> together with the right of ingress and egress to and from said pipeline or pipelines, or any of them, over and across said lands and adjacent lands of Grantors with the further right to maintain the right of way and easement herein granted clear of trees, undergrowth and brush to the extent Grantee deems necessary to exercise the rights granted herein.
>
> Should more than one pipeline be laid under this grant, at any time, an additional consideration, calculated on the basis of $5.00 per lineal rod, shall be paid for each line after the first line. It is agreed that all of said pipelines shall be located within a strip of land fifty (50) feet in width, the center line of which shall be the center line of the first pipeline hereafter installed by Grantee across, over, through and under said lands.
>
> Grantors shall have the right to use and enjoy the above described premises, except as to the rights herein granted; and Grantors agree not to build, create, or construct, nor permit to be built, created, or constructed, any obstruction, build-

ing, lake, engineering works, or other structure over said pipeline or pipelines. Grantee hereby agrees to pay any damages which may arise to growing crops, timber, fences, or buildings of said Grantors from the exercise of the rights herein granted; provided however that after the first of said pipelines has been laid Grantee shall not be liable for damages caused on the right of way by keeping said right of way and easement clear of trees, undergrowth and brush in the exercise of said rights. Said damages, if not mutually agreed upon, to be ascertained and determined by three disinterested persons, one thereof to be appointed by Grantors, one by Grantee, and the third by the two so appointed, and the written award of such three persons shall be final and conclusive.

It is agreed that any payment due hereunder may be made direct to said Grantors or any one of them.

Any pipeline or pipelines constructed by said Grantee across lands under cultivation shall, at the time of construction thereof, be buried to such depth as will not interfere with such cultivation.

The rights herein granted may be assigned in whole or in part.

The terms, conditions, and provisions of this contract shall extend to and be binding upon the heirs, executors, administrators, personal representatives, successors and assigns of the parties hereto.

TO HAVE AND TO HOLD said rights and right of way, easement, estates, and privileges unto the said MNO Pipeline Company, its successors and assigns, so long as said right of way and easement are used for the purposes granted herein.

IN WITNESS WHEREOF, the Grantors herein have hereunto set their hands and seals this the _____ day of _____, 1975.

He also finds that there are significant omissions, and he prepares the following paragraph to be added:

Should the Grantee fail to begin construction of the said pipeline on the right of way herein granted within one (1) year from the date hereof, or should the Grantee discontinue use of the same at any time for a continuous period of one (1) year, all rights and title herein granted and conveyed shall automatically revert to Grantors, their successors or assigns.

Since damages will be settled in advance, the buyer also submits a release for Bryant to sign. The lawyer approves it after noting that it covers only damages on the ROW strip itself and not damages that might be done elsewhere.

Bryant gives these revisions to the buyer; with some reluctance, the company agrees; a new contract is prepared and executed; the sale is closed. Bryant notes the action on his status map. Salvage of the timber is now the job of MNO, and Bryant instructs the forester to inspect the area after construction is complete to see that no damage is done outside the ROW. His CPA tells him that he will not have to pay any tax on the gain realized if he can invest the proceeds in similar property within 12 months, and both Bryant and his forester begin looking for a desirable tract.

CLEARCUT AND PLANT

At the 1975 review conference, the forester reports that there have been no uncontrolled fires since the 1973 grazing lease was granted and suggests a plan for SW¼, Section 26, and that part of Section 35 north of Highway 42. This area contains the pasture relinquished by Rice in 1972 and patches of trees suitable for poles, piling, sawtimber, and pulpwood but too scattered to utilize available growing space. In spite of the absence of fire, natural reproduction in the timbered area has been unsuccessful, and the old pasture is too large an opening for natural seeding. The forester estimates that 25% of the total area of 200 acres is idle. To solve these problems and to provide timber income, he suggests that the entire area be clearcut, burned, treated with TSI work, and planted. He recommends that planting crews utilize natural reproduction wherever possible and use a 9 ft. × 9 ft. spacing with close attention to care and spacing of seedlings. He estimates total sale volume at $20,000 and total cost of burning, TSI work, planting, and supervision at $4000. Considering all facets as usual, Bryant approves the plan, provided all income and expenses will fall in 1976.

At the time of the 1972 pulpwood thinning, the CPA pointed out that pay-as-you-go sales were the only way to *guarantee* that the income would be taxed as capital gains, but he also said that lump-sum sales *might* receive the same treatment. The forester reports that, under normal conditions, the lump-sum method produces 50 to 100% more money. In a sale like the one planned, he says the difference could be much larger; if a pay-as-you-go buyer processed pole-type trees into pulpwood, the stumpage return would be greatly reduced. Bryant selects the lump-sum method since it will likely produce more money net to him even if it does not receive capital gains treatment.

During field work for the sale, the forester modifies the plan by moving the boundary of the sale area slightly west so that it follows the stream. Using the procedure described on pages 181 to 184, he solicits sealed bids in early December 1975. The sale announcement states that cutting shall begin not earlier than January 1, 1976 and end not later than November 30, 1976. The deed form to be attached is much simpler. For the first three paragraphs of the one shown on pages 182 to 184, he substitutes the following:

For and in consideration of the sum of One Hundred Dollars ($100.00), cash in hand paid, and other good and valuable considerations, the receipt and sufficiency of all of which are hereby acknowledged, Robert Bryant, hereinafter called "SELLER," does hereby convey and warrant unto _____, hereinafter called "PURCHASER," all merchantable timber on the following described lands:

(enter description here) Purchaser shall have the right of ingress and egress on, across, and over the lands owned by Seller for the purpose of logging the timber conveyed herein.

Remainder of the deed is the same as shown earlier.

Seven bids are received, and Bryant accepts the highest. The lawyer prepares the conveyance, and the forester closes the sale on January 1. Cutting is completed on schedule. A contractor (the same one used in 1967) performs the necessary TSI work, burning, and planting under supervision of the forester; the entire job is complete by late December. Bryant plots this work on his status map. Although several different operations have occurred on the same area, the most important aspect is the final condition; therefore, he uses only a symbol for planting to record it.

SALE OF LAND

In mid-November 1976, H. E. Jones, Bryant's neighbor by the southeastern portion, desires to enlarge and consolidate his farm and offers to buy 150 acres described as E½ of SE¼, Section 34, SW¼ of SW¼, Section 35, and NW¼ of NW¼ less West 10 acres, Section 2. Jones offers $200.00 per acre, provided Bryant will convey some minerals and will allow him to pay for the land over a period of years. Bryant tells him that the area is now leased to Rice through December 31, that the trade cannot be effective until after that date, and that he needs a few days to study Jones's offer. Jones agrees.

Bryant then studies the consequences of this sale, and taxes play a prominent role. Jones wants some mineral interest, and Bryant decides that the most he is willing to convey is one-eighth. What is his depletion basis? It is (150 acres × $10.00 = $1500.00) + (150 acres × ⅛ × $5.00 = $93.75) = $1593.75. Subtraction of this from the price of $30,000.00 leaves a capital gain of $28,406.25, which is subject to a 4% state tax, ($1136.25) and a federal tax of 25% ($7101.56), leaving net proceeds of $21,762.19. He can invest this sum at 6% to produce an annual income of $1305.73.

Then he determines his present income from the 150 acres. Rice pays an annual rental of $900.00, but his lease denies Bryant use of 70 acres of cultivated fields, 15 acres of pasture in NE¼ of SE¼, Section 34, two houses, and a barn. Annual ad valorem taxes on these are $196.63, so net income from the lease is $703.37. Both sums represent ordinary income subject to combined taxes of 44%; reducing them by this percentage leaves $731.20 versus $393.89. The sale begins to appear attractive.

At Bryant's request, the forester inspects the proposed sale area and reports that it contains merchantable timber worth $6000.00. Bryant can cut this now or allow it to grow; the forester estimates that it will produce about $360.00 to $400.00 annually. Although these figures must be reduced by income taxes and selling expenses, they are an argument for keeping the land. The 65 acres of the sale area used by Bryant to grow timber cause an annual tax burden of $63.38. Additional expense is caused by boundary-line maintenance; the sale area has two miles of boundary line. Nevertheless, these two expenses are not large enough to offset estimated growth.

Bryant asks the CPA to check his calculations and comment on other aspects of the sale, and the CPA points out two advantages of an instalment sale. First, federal income taxes are graduated, and each advance into a higher bracket causes a higher tax rate. If Bryant's regular income remains unchanged, he can receive additional income of about $2000.00 annually without moving into a higher bracket. A capital gain of $28,406.25 will increase his taxable income by $14,203.13, over $12,000.00 of which will be taxed at higher rates. The CPA states that Bryant can save money on taxes by spreading his gain over several years and recommends that he make the sale for 20% down and 20% in each of four following years, with interest of 8% on the unpaid balance. Second, he suggests that this procedure will allow Bryant to keep his capital invested at all times. Bryant may not be able to find a desirable investment for the entire sum immediately, and he will probably be able to sell the Jones note if an outstanding opportunity comes along later.

Considering all these items, Bryant decides to make the sale, provided he can raise the price to $250.00 per acre ($37,500.00), which would compensate him for merchantable timber on the land and its growth value. He tells Jones that he will sell at this price, convey one-eighth of the minerals, and accept 20% of the sale price on December 31, 1976, and 20% on December 31 of each of the four following years, with interest at 8% on the unpaid balance. Jones insists on one-quarter of the minerals. Bryant, knowing that he has excellent title information, says that he will provide an up-to-date title opinion and pay all closing costs if Jones will accept one-eighth of the minerals. (This detail illustrates the value of preparation; Bryant offers a concession that costs him little and retains an extra one-eighth of the minerals.) They finally agree and meet in the office of Bryant's lawyer the following morning.

After describing the agreements they have already reached, they settle other details brought up by the lawyer. Bryant is to pay ad valorem taxes for 1976 and buy mineral documentary stamps. At Bryant's expense, the lawyer is to prepare an up-to-date title opinion and all necessary papers, give these to Jones or his lawyer by December 26, serve as closing agent, seeing that all documents are properly recorded, and transfer the assessment from Bryant to Jones. They select December 31, 1976, as the closing date and agree to meet at the same place. Bryant notifies Rice that his lease is terminated. To avoid misunderstandings, the lawyer prepares, and Jones and Bryant execute, the following purchase and sale contract:

---

25 November 1976

Mr. H. E. Jones
Box 189
Anytown, State

Dear Mr. Jones:

This letter will serve as a memorandum of the agreement reached between us in the office of Mr. Edward Weems [the lawyer] today.

I agree to convey to you by General Warranty Deed, subject to the exceptions hereafter noted, the land described as E½ of SE¼, Section 34; SW¼ of SW¼, Section 35, all in Township 5 North, Range 10 West, and NW¼ of NW¼ less West 10 acres, Section 2, Township 4 North, Range 10 West, for which you agree to pay me $37,500.00, subject to approval of the title by your attorney. I will convey to you only an undivided one-eighth interest in the oil, gas, and other minerals; the thirty acres located in Section 2 are already leased to Gilchrist Oil Company, and the conveyance will be subject to this lease.

One-fifth of the purchase price will be paid at closing. The balance will be represented by a promissory note executed by you and secured by a mortgage on the land conveyed. The promissory note shall be payable in four equal annual instalments of principal commencing 31 December 1977 and December 31st of each year thereafter plus interest at the rate of 8% per annum on the unpaid balance until paid in full.

I agree to pay all ad valorem taxes for 1976 as well as the cost of all documentary and mineral stamp taxes, title opinion, and closing cost. You will pay ad valorem taxes for the year 1977.

Mr. Weems will make an examination of the title, prepare title opinion, deed, note, and mortgage, and furnish you with all of this by December 26th. We will meet in his office on 31 December 1976 to close this transaction. Mr. Weems is to record all instruments and furnish the necessary proof of the transaction to Gilchrist Oil Company and the county tax assessor.

If this expresses your understanding of our agreement, please indicate your acceptance below.

Yours truly,
ROBERT BRYANT

ACCEPTED:

_____
H. E. Jones

After completing the title examination, the lawyer incorporates the agreements of the purchase and sale contract into the forms required by state law. Jones and his lawyer approve everything; the sale is closed on schedule; the closing attorney performs his functions; Bryant notes the action on his status map.

BOUNDARY-LINE MAINTENANCE

As soon as the sale to Jones is definite, Bryant must change his boundary lines to exclude this portion, and the status map shows that boundary-line maintenance is overdue. The forester reports that the paint put up by the surveyor in 1965 and 1967 is visible but dim and should be renewed. Excluding the portion sold to Jones and that maintained by the XYZ Paper Company, there are only six miles of boundary line, and the forester quotes a price of $100 per mile for this work. Bryant orders this done, receives the forester's written report when it is complete, and plots the action on his status map.

Bryant has now plotted almost the entire history of his tree farm for over eleven years on his map, and, although it is becoming cluttered, it is still legible and useful. On page 204, it is shown as it appears on January 15, 1977.

BRYANT'S SYSTEM AND THE LARGE INDUSTRIAL TREE FARMER

Bryant's system has value for all tree farmers; it is almost essential for the large industrial tree farmer because of turnover in personnel, complexity of forest management, and the increased amount of information necessary. The larger the area owned, the more fragmented the records usually become. Accounting records are often kept in one department, legal records in another, and forest-management records in a third, and it becomes increasingly difficult for the manager to see all facets of his problem. Unfortunately, the need for clear vision is even more pressing because of the large capital sums involved. Each company has its special problems and solutions to these problems, but two of Bryant's devices can be adapted for all.

First is his status map. A large company ownership should be divided into parcels that can be conveniently shown on 8½ in. × 11 in. sheets. Each district forester (the man directly responsible for field management, regardless of his title) should have a set of maps covering his land; if no other method is available, be can make them with tracing paper and pencil. He should be required to keep them up-to-date and as comprehensive as possible. The more information they contain, the more valuable they are to his company.

He should bind them in a file folder or any durable binder that will allow them to be taken readily into the field at any time and used continuously. Failure to maintain them properly should be grounds for discipline. Many foresters are reluctant to keep records and prefer the satisfactions of fieldwork; nevertheless, the keeping of these records is a must for continued company prosperity. At least one additional copy should be kept at company headquarters and revised yearly or at some other convenient interval. Many copying machines are now available, which, by eliminating tedious manual work, reduce to a minimum the labor of keeping additional copies. Finally, the district forester should be required to turn these records over to his successor when he is promoted or transferred or when he leaves the company.

The maps should show all forest-management practices important to the company. Names and addresses of neighbors are always useful. Government field notes on boundary lines are usually not necessary, but dates and kinds of maintenance and information about variations from true locations of lines and corners are desirable. Rights of way, encroachments, unproductive areas such as old beaver ponds and fences, and similar items one step removed from forestry must be plotted.

The second valuable part of Bryant's procedure is his annual review. He is stockholder, board of directors, and executive officer in his enterprise, and annual reviews and reports are common at higher levels of large companies. They are also desirable for lower echelons. The end of the calendar or fiscal year is a good time for review, since other departments are getting figures together on the same period, but some definite date for review of forestry is essential for efficient operation. In the life of a forester, one day moves quickly into another, and it is often easy to lose sight of major goals in the day-to-day administration of forests. Progress reports, reappraisals, and critiques measure the success of the enterprise and help keep all eyes on the main objective, and progress of the past year or other period is an excellent guide for plans for the future.

FINANCIAL RECORDS

Bryant's financial records are simple but effective. Just prior to his purchase, he opened a special bank account for the property. He paid for the property with a check drawn on this account, and he deposits all income in, and pays all bills from, the same account. He puts a brief explanation on each check stub and deposit slip and keeps his checkbook as a permanent record.

Each year, he opens a special file for receipts and expenditures in connection with the property. Into it he puts deposit slips, bank statements, canceled checks, tax receipts, bills from advisers and others rendering services,

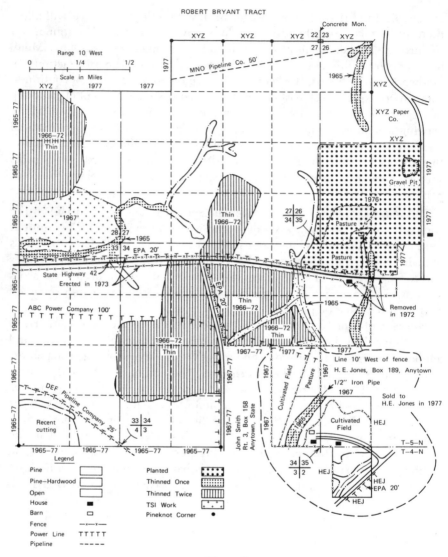

Figure 5

memoranda of out-of-pocket expenses and trips he takes to administer the land, and so on. At the end of the year, his checkbook and this file are sufficient to prepare his profit and loss statement and income tax return, and he has evidence to substantiate each item. Each December 31, he seals the file and stores it. Everything in it is of transitory importance and does not need to be kept longer than the statutory period for all financial records. Everything in the main file on the property, that containing the forest-management records and so forth, has a bearing on his title and is of permanent importance. Therefore, some separation is necessary to keep the main file down to manageable size.

As you can see from the history of his operations, his transactions are infrequent, usually no more than two or three receipts and 10 or 12 disbursements; therefore, separate books of account are not needed. Even such an important item as depletion basis can be determined from his checkbook alone. The stub of the check for the purchase price contains the allocation into various components, and the checkbook shows all later transactions in detail. Twenty-five years after purchase, it is a simple matter to calculate his depletion basis. If your property is larger, however, you have more frequent transactions and probably need more elaborate records; your tax adviser can design them.

CONCLUSION

After three partial cuts and one liquidation, one land sale, one ROW sale, and 12 years of growth, Bryant needs another timber inventory and growth study. He can no longer make accurate financial predictions, and, although he does not desire annual incomes, he wants to compare his tree farm with other available investments. Fortunately, however, many important facts gathered in the original inventory are now part of his permanent records. He knows there is no merchantable timber on the area planted in 1976, so there is no need for inventory of this, and he has a good idea of stand conditions and the location of operable volumes. The only missing links are timber volume and growth rate, and these may be determined for the tract as a whole by a 5% inventory at modest cost. Therefore, good records have another advantage; they reduce the cost of future inventories.

This has been the story of Robert Bryant, covering a period of 25 years. His methods were not perfect; his results leave room for improvement. Nevertheless, through 1977 and beyond, he has found tree farming a rewarding enterprise, and I hope his example helps you reach the same goal. Good luck.

Glossary

Tree farming has its own terminology, and we need to define some terms to make sure we know what we are discussing. Sometimes a bare definition is not sufficient. I suggest that you read these definitions and then refer to them from time to time as necessary.

BILTMORE STICK—a graduated stick about 3 ft. long used mostly to make rough measurements of tree diameters quickly. The user holds it against the tree and a specific distance from his eye, lines up one end with one side of the tree, and reads diameter from the graduation where other side intersects the stick. Improper use can cause large errors in volume estimates.

CUTTING CYCLE—the frequency of logging operations on the same area, expressed in years. Let us assume that you own a tract that contains trees of all sizes and ages. If you have decided to cut this tract every 10 years to thin the small trees and harvest the large ones, you have chosen a 10-year cutting cycle.

DBH AND DBH CLASSES—the abbreviation DBH means diameter breast high (4.5 ft. above ground level) outside the bark and is expressed in inches. Careless speakers often use the term to mean DBH class, but it is necessary to know the difference. Trees are divided into 1-in. and 2-in. DBH classes for convenience. With a 1-in. class, a 10-in. DBH tree is more than 9.5 in. DBH and less than 10.5 in. DBH. With a 2-in. class, a 10-in. DBH tree is more than 9.0 in. DBH and less than 11.0 in. DBH. Keep this in mind; you will avoid many arguments.

DELIVERED PRICE—the sum paid for the tree at mill or railroad.

EASEMENT—the right or privilege of making limited use of another's property.

EMINENT DOMAIN—the power to take private property for public use. National or state legislatures may pass this power to agencies that serve the public, such as pipeline companies, highway departments, power companies, and so forth.

HARDWOOD—a loose term generally including all species of trees that lose their leaves in winter. Some hardwoods such as magnolia retain leaves throughout the year; other species such as larch and bald cypress lose their leaves but are not hardwoods. Hardwoods do not bear cones, softwoods do. Any forester can explain what the term includes in your area.

INCREMENT BORER—a small tool used by foresters to determine growth of a tree. It is a hollow tube with a handle on one end and a spiral cutting bit on the other. When pressed against the bark and turned by hand, it moves into the tree toward the heart and cuts a section slightly smaller than a pencil. This section, called an increment core, can be removed from the hollow tube with an extractor, which is also part of the tool. With most species, the increment core shows each annual ring clearly and allows you to determine how fast the tree is growing. Damage to the tree is insignificant compared with the value of the information obtained.

LOGGING COST—the sum necessary to move the tree from the woods to a mill or railroad.

MBF—a convenient way to write "thousand board feet."

MERCHANTABLE TIMBER—any timber that can be sold. Merchantable trees usually have a minimum DBH of 5 in., but certain products such as fence posts may come from smaller trees.

POLES AND PILING—particularly straight trees that meet exacting requirements so far as diameter and taper are concerned can be manufactured into poles and piling. In general, piling comes from sawtimber trees and poles from both sawtimber and pulpwood trees. Trees processed into poles and piling often bring 50% more stumpage than when used for other purposes. The danger in special cuts for poles and piling is that you may not find a buyer for the remaining timber.

POSSESSION, ADVERSE AND OTHERWISE—see discussion beginning on page 61.

PULPWOOD TREES—trees over 5 in. DBH that are unsuitable for sawtimber because of size, crook, or other defect. Trees cannot be classified as pulpwood on the basis of size alone.

PULPWOOD VOLUME MEASURES—see discussion on page 72.

REPRODUCTION—well-established seedlings from about knee height up to pulpwood size. Occasionally they are separated into DBH or age classes.

ROTATION—the period of years required to reproduce, grow, and harvest stands of timber in order best to accomplish definite objectives of management. The final crop of a pulpwood rotation is pulpwood; the final crop of a sawtimber rotation is sawtimber; and the second type of rotation is about twice as long as the first in time. There are other technical forestry aspects to defining this term, but they are not necessary for your purposes.

SAWTIMBER TREES—usually trees 11 in. DBH and up, but definition of the term varies with species and geography. They must be reasonably straight and free of defects caused by fungi, insects, fire, or the way in which the tree developed. Sawtimber trees may be manufactured into lumber, veneer logs, poles, or piling; the term is only a general one that primarily designates a major division by size. Definition of the term may also vary with changes in economics. You can saw a two-by-four from a small tree, but logging of small trees is expensive and is usually practical only during strong lumber markets. As prices decline, lumber companies often raise the minimum diameter of the trees they will buy.

SAWTIMBER VOLUME MEASURES—see discussion on page 73.

SEED-TREE CUT—removal of all merchantable timber except those trees necessary to produce seed for the next crop.

SLASH—tops and limbs left in the woods after cutting operations.

SOFTWOOD—all species that are not hardwoods.

STAGNATION—as the term is commonly used in tree farming, *stagnation* occurs when too many trees are growing on the same area. In the fierce fight for water, light, and food, these trees are barely able to stay alive. Since the productive capacity of the soil must be divided among so many individuals, growth appears to stop, but it is only hard to see because it is so widely distributed. You can understand what this does to your annual income. The good tree farmer tries to prevent stagnation by proper planning, or to relieve it by thinning as soon as possible.

STATUTE OF LIMITATIONS—see discussion on page 118.

STUMP DIAMETER—usually, the average diameter outside the bark measured 12 in. above ground level on the high side of the tree. A rule of thumb says that the diameter inside the bark at this point equals DBH; remember that this is only a rule of thumb. Stump diameter is often used in timber sales, and you do not encounter diameter classes in this case. A conveyance should specify the height above the ground, whether inside or outside the bark, and how to determine average diameter.

STUMPAGE PRICE—the sum paid the owner for the tree as it stands on its stump.

SUPPRESSED TREES—those that have nearly lost the battle for survival. Overtopped by their neighbors and reduced to a pole with only a skimpy crown, they may die within a short time and have usually lost the ability to resume normal growth if released.

THINNING—partial cutting where the material removed is sold and produces some return. It is done to prevent or relieve stagnation and to concentrate growth on better trees in the stand. *Precommercial* thinning is done before the trees that are cut are large enough to be sold, and it may be included under TSI work. Although it is nearly always a deductible expense, the cost is apt to be prohibitive even when you use machines designed for the purpose. An ounce of prevention is worth several pounds of cure.

TIMBER CRUISE—means the same as *timber inventory*. The term probably arose because the man performing the work navigates a path through the property much as a ship's captain follows his path on a cruise.

TRESPASS—as a management problem in tree farming, trespass generally means theft or unauthorized cutting of timber.

TSI—stands for timber stand improvement and means almost anything done to improve the condition of a timber stand without producing revenue at the same time. Pruning and removal of worthless trees that impede the growth of desirable trees are examples.

Index